CONSULTING

the GENIUS

of the PLACE

CONSULTING

the GENIUS

of the PLACE

AN ECOLOGICAL APPROACH TO A NEW AGRICULTURE

WES JACKSON

COUNTERPOINT

BERKELEY

Library of Congress Cataloging-in-Publication Data

Jackson, Wes.
  Consulting the genius of the place : an ecological approach to a new agriculture / by Wes Jackson.
    p. cm.
  ISBN 978-1-58243-780-4
  1. Agriculture—United States. 2. Agricultural ecology—United States. 3. Sustainable agriculture—United States. 4. Food supply—Environmental aspects—United States. I. Title.

S441.J247 2010
577.5'50973—dc22

                              2010015641

Interior design by Amber Pirker
Jacket design by Jeff Clark

COUNTERPOINT
2560 Ninth Street, Suite 318
Berkeley, CA 94710
www.counterpointpress.com

Printed in the United States of America
Distributed by Publishers Group West

10 9 8 7 6

To Joan

CONTENTS

PREFACE

The poet Alexander Pope in 1731 coined the phrase that inspired the title of this book. Pope's principle had great impact on English landscape gardening. Any landscape architect who did not converse with the local genius would be castigated by the client for his ignorance of the "genius of the place."

Pope was not thinking of vast agricultural landscapes, certainly no scale approaching the corn and soybean fields of Iowa or the wheat and sorghum fields of Kansas. Though Pope was thinking about gardens in his admonition, why not scale up, and extend his idea to the agricultural lands in our ecosphere? Why not give ourselves the opportunity to have these ancient-beyond-memory ecosystems teach us how to be anywhere on earth? Since agriculture began we have needed corrective action but never more than now. The explosion of power in the hands of individuals is at its height coupled with our still-exploding numbers. The deficit spending of the earth's capital since we started agriculture ten thousand years ago has greatly accelerated in our time. The only good examples of an alternative way are among nature's ecosystems that feature material recycling and run on contemporary sunlight.

We have to think beyond the ecosystem to the ecosphere itself. So, when this book has been read, closed, and reflected upon, I hope the reader will have concluded that the idea that humans live within a "life support system" is not the best way to think about our relationship, that a more correct idea is that all organisms, including us, are embedded within a living ecosphere, a supraorganism, not superorganism. Moreover—and I hope the reader is ready for this—it is the only truly creative force at work in the world. The scientist at the bench or the artist at the easel is only creative in the context of a civilization, whose scaffolding is constantly in need of repair and replacement from the capital stock of the ecosphere, which has been drawn down and the sinks increasingly filled (like the atmosphere) since the early days of agriculture. Stated otherwise, civilization's demands on the ecosphere's resources to repair the scaffolding are greater than the renewal rate the ecosphere provides, and as the sinks (oceans, atmosphere, and more) reveal their limits for supporting the current biota, source-sink becomes one subject.

How to get a grip on this? For starters, I hope that a larger definition of life that includes the physical world can overcome the conflation of life and organism that limits our perspective. I am not advancing the necessity to believe in spooks. Indeed, I am confident that a prior commitment to materialism need not detour one from acknowledging that

PREFACE

the earth lives. I don't believe that humans can destroy the earth's capacity for renewal. I do believe that full recovery from our abuse will come in geologic time (millions of years), not human time (200,000 years), certainly not agricultural time (10,000 years). We want to stay here, but we want all the other species embedded in the same ecosphere within their ecosystem to stay with us, too.

Here is a final hope:

As our minds sweep over the past and back to the present, I want them to center on the natural ecosystems still with us as our primary teachers. They are our source of hope. Reduced in number and limited in scale, they still hold answers to countless questions we have not yet learned to ask. The best place to begin to apply the knowledge we have learned and will learn is in agriculture.

xi

PART I

Some History and Assumptions

CHAPTER ONE

INTRODUCTION

My wife, Joan, and I try not to interrupt one another when we are reading and having our coffee in the early morning. We are usually successful, unless an arresting piece of information or idea presents itself. One morning as she was reading *The History of Love*, a novel by Nicole Krauss, she interrupted with "Listen to this." The protagonist in the story, a young man, describes a woman in his village in Poland who had paid special attention to his writings. It was when Hitler's troops had entered Poland, and for whatever reason, this woman had moved from their village. Joan read aloud the following passage:

> After she left, everything fell apart. No Jew was safe. There were rumors of unfathomable things, and because we couldn't

fathom them we failed to believe them, until we had no choice and it was too late.

We both fell silent. We knew what the other was thinking.

Here is the modern problem: we continue to hear "unfathomable things." Beyond the climate scientist's statements that the earth is warming up, the modelers say we can expect more severe storms. The permafrost is melting five times faster than earlier predictions. Rapid fossil fuel consumption, rapid population growth. Worldwide degrading of soils, loss of biodiversity, disruption of ecosystems. We have somewhere between 150 and 1,000 dead zones in our oceans from nitrogen applied to agricultural fields. The list of "unfathomable things" associated with climate change alone seems endless.

Too many reputable scientists publishing in refereed journals are taking the discussion beyond the level of mere hearsay for us to ignore. I am one who trusts the conclusions drawn by the National Academy of Sciences and the Intergovernmental Panel on Climate Change and the consensus they handed down: human-induced climate change is here.

No matter that there are naysayers on all negative statements about the state of the ecosphere. One of the first lessons a scientist learns is that 100 percent agreement on scientific matters is not to be expected. There may be a consensus, but most know a consensus can be reversed. But that reversal won't come from talk radio or TV pundits. It will come from those who publish under the stringent demands of scientific journals.

It is scientists' right to disagree with the consensus in their field. I could even argue that it sometimes is their obligation. I'm told that there is a small number of medical researchers who do not believe the HIV virus is the cause of AIDS. One is on the faculty at the University of California at Berkeley. I had a friend, on the faculty before he died, who knew him and who thought his argument plausible. If the researcher wants to overturn the dominant idea with data, power to him.

However, I don't yet want a blood transfusion from someone who authorities tell me is carrying the HIV virus. I'll stick with the overwhelming consensus of most scientists with little question so long as it is outside my field of training.

I'm a geneticist, not a climatologist, and have no training in that discipline. But, because I hold to a scientific way of knowing, I accept the view of 99 percent of the men and women who publish in refereed journals on climate. What have they concluded? That humans burning fossil fuels—coal, oil, natural gas—are the major source of the carbon dioxide increase in the earth's atmosphere. This is causing the planet to heat up, and for our health, and the planet's, we must cut our fossil fuel consumption to 20 percent of what it is now long before century's end.

Unfortunately, for purported balance, the media give disproportionate time to the 1 percent who are not convinced. But there are thousands of scientists involving 130 countries associated with the Intergovernmental Panel on Climate Change plus the National Academy of Sciences, and among them the consensus is overwhelming.

I used to ask my students if they believed that the earth went around the sun or that the sun went around the earth. Of course, they all believed the former. I would then ask them why they believed that and if they ever stopped to check it out for themselves. They hadn't. Few people have in our time.

How do I know the pancreas secretes insulin, that it isn't the liver or the spleen or my femurs? How do I know that grinding plates of the earth cause earthquakes? That fire feeds on oxygen? These are only a few of the countless conclusions we mostly accept because others devoted to a scientific way of knowing, in their particular fields, tell us it is so.

The question finally becomes, what is prudent? If 99 percent of climate scientists turn out to be wrong, what has been lost? If we act on the consensus and reduce fossil fuel consumption and turn out to be wrong, what is lost? But if the consensus is right and we refuse to respond, climate change is likely to present the greatest challenge to public health in the history of our species.

The primary message of this book is about saving the ecological capital necessary for an assured food supply into a distant future without fossil fuels. Once again we can look to science as a way of knowing.

By doing so we will draw on much of the same spirit of that scientific revolution that weakened long-held beliefs and superstitions. This attitude, this way of thinking, arose in Western civilization beginning in the 1300s. The changed attitude came from many sources. The age of voyages increased our imagination. So did a rediscovery of the thinking of the ancients. Then there was the Reformation and the thinking of the fathers of modern science on both sides of 1600—men like Copernicus, Galileo, Francis Bacon, and René Descartes, followed by Sir Isaac Newton. These men are our intellectual ancestors. Out of their thoughts and approach to knowing came a nonreligious literature. Not only do we live longer, this way of knowing helped us understand our biological and even our cosmic origins. (There are problems with a total reliance on this way of thinking, as will become apparent in what follows.) I want this scientific way of knowing kept alive. But that is not enough for our time. Civic action among scientists, themselves, activism beyond our professions, will be necessary immediately, for it is not just climate change that is in the realm of the unfathomable at issue here. The speed of change is accelerating and is already changing the rules of behavior. Consider, for example, that the twenty-two-year-old has lived through over half of all the oil ever burned; the ten-year-old, a quarter. In my first twenty-two years (1936–1958), I lived through only 16 percent as much oil consumption as has a twenty-two-year-old living today. Another speed indicator is the rate of the world's population growth, which has tripled in a single lifetime, doubled since John Kennedy was president. We stand at a moment in history unlike any other. Anyone who had died by 1930 never lived during a doubling of the human population. According to biologist Joel E. Cohen, anyone born after 2050 likely won't either. This 120-year period in population growth, by no mere coincidence, has come accompanied by a comparable increase in fossil fuel consumption of which the oil just mentioned is an example.

The entire discussion brings us back to a major consideration of this book: feeding humanity in a world where the fossil fuel resources standing behind agriculture are steep in decline. The problem is even worse than that. The 2005 Millennium Ecosystem Assessment synthesis report says that "over the past 50 years, humans have changed ecosystems more rapidly and extensively than in any comparable period of time in human history, largely to meet rapidly growing demands for food, fresh water, timber, fiber and fuel. This has resulted in a substantial and largely irreversible loss in the diversity of life on Earth." Countless specialists tell us that the disruption to biodiversity and ecosystems is mostly due to agriculture. And these specialists conclude that the degradation of ecosystems could grow significantly worse during the first half of this century.

As will become apparent before the end of this book, it is my view that the future of agriculture long before the end of the fossil fuel interlude will depend on knowledge gained from our ecosphere's wild ecosystems. To lose them in the interest of agricultural expansion now would be akin to destroying all the schools and libraries for heat and materials now.

Scientists Mathis Wackernagel, et al., publishing in the *Proceedings of the National Academy of Sciences* in July 2002, had said it well:

> Sustainability requires living within the regenerative capacity of the biosphere. In an attempt to measure the extent to which humanity satisfies this requirement, we use existing data to translate human demand on the environment into the area required for the production of food and other goods, together with the absorption of wastes. Our accounts indicate that human demand may well have exceeded the biosphere's regenerative capacity since the 1980s. According to this preliminary and exploratory assessment, humanity's load corresponded to 70 percent of the capacity of the global biosphere in 1961, and grew to 120 percent in 1999.

The warnings about our assault on the ecological capital behind food production are not new. As early as 1995, Cornell University scientists David Pimentel, et al., said, "Soil erosion is a major environmental threat to the sustainability and productive capacity of agriculture." In the four decades prior to 1995, the scientists found that "nearly one-third of the world's arable land [has] been lost by erosion at a rate of more than 10 million hectares per year." That is twenty-four million acres, half the size of Kansas. The authors further estimated that an investment of over $6 billion per year on cropland alone would be necessary to reduce annual U.S. erosion rates from seventeen tons per hectare to what they called a sustainable rate of about one ton per hectare. That was in 1995. Some have called Pimentel's conclusions too extreme. But does it matter? Even if erosion is one-tenth as bad and the cost proportionally less expensive, we are losing the elements of which we are made. But if so, how off the mark are they?

The Conservation Reserve Program pulled from production more than thirty-six million acres in the first round. By planting perennials the overall loss of soil greatly declined by about 40 percent. There has also been a reduction due to minimum-till and no-till methods, but in those cases they feature the herbicide, and now recent resistance to the herbicide is on the increase. Even here, because the crops are annuals, they do a poor job of managing the nutrients, especially the nitrogen, which slips seaward every year. But even with the reduction in rate, it still amounts to deficit spending. And besides, on a global basis, soil erosion is accelerating.

There are thousands of environmental issues that need to be addressed here and there throughout our ecosphere, but at the moment it is climate change that seems to be our wake-up call and perhaps sums up all of our other problems best. Melting glaciers, climbing carbon dioxide levels from 260 parts per million (ppm) in 1840 to over 360 ppm in 2010. Unless we respond quickly, by 2050 this number is expected to increase to 500 ppm and by 2100, to 750 ppm. Modelers have predicted temperatures rising anywhere from 3 to 7.5 degrees Celsius. At the depth of the ice age, average temperatures were only 5 degrees cooler than now.

From the results of the warming already under way, the distribution of plants and animals on land and in water has been and continues to be altered. The migrations and breeding patterns of birds are changing. The flowering and fruiting of plants are already being disrupted. Buds are bursting sooner, flowers are blooming earlier. Animals break hibernation sooner, migrate earlier. As Camille Parmesan warns, "Established predator-prey relationships as well as insect-plant systems have been disrupted." The most restricted species ranges, which tend to be mountaintops and at the poles, are experiencing the most severe contractions. Tropical coral reefs and amphibians are dying. Plant and animal ranges are moving poleward. Warm-adapted communities on all continents are expanding. Parasites and the organisms that carry them have influenced human disease dynamics.

The list of "unfathomable things" goes on and on. Is there any good news not overshadowed by these "terrible truths"? In an odd way, yes or maybe. This time the insight comes out of the literary tradition from the poets. A few years ago, the poet and scholar Kathleen Raine, commenting on the works of T. S. Eliot, said, "Eliot has shown us what the world is very apt to forget, that the statement of a terrible truth has a kind of healing power. In his stern vision of the hell that lies about us . . . there is a quality of grave consolation. In his statement of the worst, Eliot has always implied the whole extent of the reality of which that worst is only one part."

What is that "whole"? And knowing that "whole," can it indeed be a source of grave consolation? We may not be able to answer the second question with certainty, but we can begin an exploration of the first.

After we have acknowledged the reality of waste, greed, and economic and population growth, we still need sources of restoration. What follows in this volume represents my perception of baseline considerations for that larger whole. I think the news, as it comes in, will continue to be worse than what we can perceive now. And as it arrives we will have to confront our most basic ignorance about limits. We will have to confront how much we should rely on technology to solve our problem. We will have to confront our attitudes toward economic growth. We will have to confront population growth and more.

Part of my journey into that larger "whole" began innocently enough. As it has unfolded over the last two decades, I have come to a conclusion, almost too bleak to state, but it does contribute to me part of that source of "grave consolation." It came from countless hours of deliberation with my colleague, the late Dr. Marty Bender.

Marty was trained in physics, chemistry, and biology, and he was a superb numbersmith. Marty directed our ten-year Sunshine Farm Study, which involved 210 acres here at The Land Institute (see color insert *fig. 2*).

The purpose of this investigation was to determine how much food could be exported through the farm gate if the sunlight that fell on that farm had to pay all of the energy bills for the operation using conventional crops. We had a biodiesel tractor and draft horses. We had photovoltaic panels to provide the electricity for the farm. We had 50 acres devoted to two five-year crop rotations plus 160 acres of grass for up to 70 head of longhorn cattle (counting the calves) at one point. We had chickens for meat and for eggs. We grew two oil-producing crops in the rotation: sunflowers and soybeans as an offset for the biodiesel tractor.

The length Marty went to determine the energy costs was mind-boggling. For example, for the biodiesel tractor he went all the way back to the mining of the ore in the Minnesota Iron Range to get an estimate of its embodied energy. Even if the farm needed to purchase a bolt, the bolt was weighed and the embodied energy recorded, including the energy cost of the trip to town to make the purchase. This meant that the pickup truck had to have a portion of its lifetime charged against the oil-producing crops we were growing. All sorts of considerations were explored. A draft animal consumes energy just standing around being a draft animal. The tractor can be shut off, but can't have a baby tractor. The tractor gives off lots of waste heat. The hen lays an egg and doesn't get a fever. The relationships among energy, information, and scale became paramount in our thinking.

It is an understatement to say that the bookkeeping alone for this ten-year study was huge. During one of those discussions as to where to draw the line, I asked Marty where he thought some of the researchers he was citing stopped in their energy assessments. His reply was telling: "When they get tired." We knew going into the project that the mental journey backward in time and across space would be impossible in an absolute sense. We weren't that dumb. What we failed to appreciate is how quickly the "scaffolding" of civilization became so elaborate and so energy inten-sive and so unknowable. For a simple mental exercise, consider taking corn from the farm to an ethanol plant. How far is the drive to the plant? That can be reasonably estimated, and so can the amount of fuel required to make the haul. But how much of the truck's manufacturing should be charged against the ethanol output? Easier than one might think. Marty handled this level with aplomb. How much of the road construction and maintenance? Now it is getting harder. These secondary and third-level considerations are somewhat manageable. But how much of the board-room meetings of the various corporations and the cost of air travel by executives and others should be charged as capital is raised to build the plant? What about the lobbyists who went to Washington, DC, or the state capital and made their rounds to the offices of the senators and congress members? How long were they stuck in traffic, engines idling? The further the loops are widened away from the farm, the more ambiguity reigns.

The "potholes of civilization" is an expression that may serve as a meta-phor for any shoring up of scaffolding likely in anyone's accounting. As our minds went over these questions, a new question arose: If we could immedi-ately remove ourselves from having access to any nonrenewable energy-rich carbon source, how much of the current standing crop of *Homo sapiens*—humans—would be here in one week, one month, one year, one decade, one century? Given the reality of material entropy, as well as energy entropy, how much of the scaffolding of civilization would be around and in what condi-tion? And what is the quality as well as quantity of energy necessary to main-tain it? More on that issue three paragraphs down.

All the deeper questions go beyond the available answers. Nevertheless, this mental journey led me to a certain conclusion. I put it in the form of a hypothesis, an unfalsifiable hypothesis (and therefore, not testable), which is:

outside a Stone Age world and before agriculture, there has not been a single human-designed product or process, including the domestication of crops and livestock, that hasn't come at the cost of the drawdown of the capital stock of the planet. Is this part of the "whole" that can become a part of the source of "grave consolation"? In fact it was for me. For what is implied is that all future scaffolding, somewhere along the line when the nonrenewables are severely scarce, will then be dependent on the powers of the renewability of our planet's ecosphere and whatever renewable technologies it can support. I realize that here is where consideration of different time frames becomes crucial. Cosmologists think in billions of years. Geologists in millions. Human time may be the length of time we have had the big brain—the 1,350-cubic-centimeter brain—which is estimated to be as low as 100,000 years and as high as 150,000 to 200,000 years. Agricultural time is the last 10,000 to 12,000 years; industrialization, the last 250 years. Historians tend to think in centuries; statesmen, a few decades into the future; politicians, until the next election. Some think of future generations not at all.

Depending on what we are discussing, renewability can vary from a fraction of one lifetime to cosmic time. When we think of the drawdown of the capital stock of the planet as the major contributor to our way of life now, we are considering the burning of fossil fuels primarily. The cutting of forests and the disturbance of soils and erosion associated with most of our agricultural output have a shorter renewable time period—maybe. It depends on where we are considering across the wide ecological mosaic of our ecosphere.

Here is one bottom line not widely considered: the net primary production capability of the biota of the ecosphere to provide humans energy-rich carbon on a renewable—which is to say, non-fossil-fuel-subsidized basis— goes down with other ecosystem services. And a corollary assumption: none of our technologies can do better than nature's renewability powers.

To believe otherwise is to cultivate an illusion based on a lack of acknowledgment of our scaffolding—developed during the industrial era—wearing out. Much of it will need high-quality energy for repair, the kind we have burned to build it, and that energy won't be around, and thank goodness, for it is a major source of greenhouse gases.

There are three possible kinds of minds at work as we contemplate the future of humanity on the only home we have had or likely will have: (1) those who accept the hypothesis as true that we cannot do better than nature (2) those who believe we can do better, and (3) those who believe that sometimes we can and sometimes we cannot do better. And "cannot" gets back to civilization being dependent on the scaffolding made possible by the five exhaustible and relatively nonrenewable carbon pools found in soil, trees, coal, oil, and natural gas.

We are forever mentioning the high population level of humans, and we should. Let's not take our eye off that ball. But we too often fail to mention the "population of things" other than people. To illustrate this, I can do no better than to draw on the insights of the economist Herman Daly, who, after humans, listed other populations, which would include cars, houses, TVs, and Pizza Huts, along with the earth's biota, such as deer, turkeys, and bobcats. What they all have in common is that they all occupy space in a finite ecosphere, they all require energy to make and repair, and they all produce waste. They dissipate. The result is low-quality heat and dispersed materials. So the energy cost of the scaffolding that stands *behind* the hybrid car or the computer wears out faster than the energy we harvest each day for the purposes we direct from the sun. What about wind machines, photovoltaic panels, and other renewables? Well, high-energy scaffolding stands behind those renewables, too. This is not to say that we should have no wind machines or photoelectric panels. But is it fair to wonder and ask, can they produce enough in their lifetime to replace themselves from scratch?

In the preface I mentioned that our creativity as artists or scientists is extremely limited compared with the creativity of the ecosphere, since all of us are dependent on civilization, which itself is dependent on the drawdown of even short-term renewable varieties of the earth's capital stock, such as soils and forests.

With such discouraging words, are we simply to surrender to a seemingly hopeless situation? I don't think so, for if such an assessment is correct, it should be obvious that a *conceptual change, followed by practice,* in the way we conduct business within our ecosphere is necessary, and, I think, possible. The conceptual change necessary bundles nearly all the problems into one and brings us to the major point of this book, which is that agriculture has the sole potential to provide the lead into a different relationship with our ecosphere. From that relationship can then come a more realistic assessment of the technology we adopt.

It won't be easy, since the dominant idea of our time, that nature is to be subdued or ignored, goes back to the beginning of agriculture. So it is an old idea. Most of us want sustainability and to be resilient. Most of us see the need for humanity to live within its means. My late friend Chuck Washburn helped me think this through, and it is my view now that if we don't get sustainability in agriculture first, it is not going to happen, and for this reason agriculture alone, ultimately, has a discipline behind it. It is the synthetic discipline of ecology–evolutionary biology. Nature's ecosystems are ancient. They are real economies. The law of return described by Sir Albert Howard operates. They can be trusted. The materials sector, the industrial sector, is recent. It has no time-honored discipline to draw on. Skyscrapers, freeways, and suburbia have been made possible due to our discovery and use of fossil carbon, not any organizing concept. Soils and forests have fed and sheltered us, but they too are in decline even with the subsidy of fossil fuels.

I have little confidence that humans can do better than nature—ever. So where is this source of "grave consolation"? If we believe that *sometimes* we can do better, our selection out of the mix of technologies will be informed by a different set of criteria for technology assessment than the mix predicated on the notion that *we cannot*. To help think this through, imagine a distance of a hundred yards on a football field, say, and that perfect sustainability or resilience is the full distance to the goal line. Imagine that five yards is where we are now. If we assume we can do better than nature, we may advance thirty or forty yards. If, on the other hand, we assume that sometimes we can and sometimes we cannot, we may go sixty yards. If we assume we cannot ever, then perhaps ninety yards. In this fallen world we will never go the full distance. We will not voluntarily go back to gathering and hunting. Whatever technology we take with us will cause us to reduce our ecosphere's carrying capacity. But if we can, during this journey, reduce the population of things—houses, cars, deep freezes, and people—we buy more time. But with that as our unachievable ideal, the time we buy could be in the thousands of years, not centuries or scores or decades. This belief that we can never do better than nature's ways may be the only source of humility for the secular mind.

René Descartes in his *Meditations on First Philosophy* advanced the idea that the world can be remade to meet our desires. The three revolutions that define the early Enlightenment period Descartes helped jump-start—scientific, political, and economic—merely validated the control of nature and the creation of economies and technologies that open up mines and wellheads. It gave license to more than hubris. It also gave license to greed and envy. "Foul is useful, fair is not," said John Maynard Keynes. Maybe he thought that once the mines and wellheads were open, then we would suspend greed and envy. We have to admit that these revolutions in science, politics, and economics have informed what has turned out to be our erroneous belief that we can remake the world to meet our interests, selfish or otherwise, without threatening consequences. All that acknowledged, the good news is that increasing numbers of humans are now awakening as we approach the end of the fossil fuel epoch.

In the foreseeable future we will be forced to power down, reduce our numbers, and cope with rapid climate change. Receiving too little attention is our ability to keep ourselves fed during this, our most important, phase of the human journey. If we are to have a soft landing, we must turn our attention to our soils and fresh water. The longer we wait, the lower the long-term carrying capacity of this still-beautiful planet will be.

Clearly, part of that "whole" behind the "terrible truths" is the economic order designed for extraction with little, if any, provision for renewal. And now for what may appear to be a bold assertion. When it comes to agriculture, it is here we have a chance to square the food-producing part of our economy with nature's economy, a real economy that features material recycling and runs on contemporary sunlight. I believe it is here that we can begin to explore the possibilities of a different economic order, this time of the renewable variety. Of course, more is on our plate than food if we are to be secure. Economics is about more than bookkeeping. It should now be clear that we, in the industrial societies in particular, have been cooking the books to falsify the deficit spending we have imposed on our ecosphere. Perhaps a new agriculture based on the way nature's ecosystems and nature's economies have worked over millennia can point the way.

The stakes are high, but maybe not as high as implied by the astronomer Fred Hoyle nearly a half century ago:

> It has often been said that, if the human species fails to make a go of it here on the earth, some other species will take over the running. In the sense of developing intelligence this is not correct. We have or soon will have, exhausted the necessary physical prerequisites so far as this planet is concerned. With coal gone, oil gone, high-grade metallic ores gone, no species however competent can make the long climb from primitive conditions to high-level technology. This is a one-shot affair. If we fail, this planetary system fails so far as intelligence is concerned. The same will be true of other planetary systems. On each of them there will be one chance and one chance only.

Chapter Two

ONE MAN'S EDUCATION

I n June 1952, the summer I became sixteen, I abandoned myself to the prairies of South Dakota to work on a ranch belonging to a childless couple. Ina, an eccentric first cousin of my mother's, and her equally eccentric Swedish immigrant husband, Andrew Swan, were in their sixties and seventies by then, and their land acreage was huge by my standards.

They lived in Mellette County, South Dakota, bordered on the south by the Rosebud Reservation and on the west by the Pine Ridge. It was here that I encountered, on a significant scale, the pitiful reality of the Native Americans and how ranchers saw them (see color insert *fig.* 1).

I was not aware of where I was either in historical time or in world-view, but on those dusty streets of White River I must have exchanged glances with older natives who, as children or young adults, witnessed the massacre at Wounded Knee. Some of them certainly had relatives at the Battle of the Little Big Horn seventy-six years earlier. This was a land of mostly native prairie. It was there that I got my first intimate engagement with a prairie landscape whose vegetative structure, or physiognomy, was determined more by its ecology than by its culture. So, even though the bison were gone, replaced by their domestic bovine relatives from Europe, the land was less disturbed by human tinkering than most landscapes appropriated for human food. Ina was Andrew's second wife; his first had been Ina's sister Bertha. Andrew and Bertha homesteaded one half section and Ina another. When Bertha died after a twenty-year marriage, Andrew and Ina married and joined their holdings and continued to add land. On some Sundays I rode horses over those prairies with those known to us as "half-breed kids." They told me of various adventures—both of their own and of their parents and grandparents—including how their Indian grandfathers had trapped eagles on this hill or that.

On Saturday afternoons, Andrew, Ina, and I would go to White River, as close as anything to a real frontier town that still existed. Some of the Rosebud Sioux would lie in the shade of the stores, and as the sun moved they would pick up their belongings and move to the shade on the other side. Out on the ranch Andrew would cuss and swear when I asked questions about the natives being removed from their land, provoking him more than once to let me know that the Indians never did anything with the land. In town, the Indian from whom Andrew and Ina were leasing some Indian land regularly charged groceries to their account. Andrew always paid, for to fail to do so meant that a neighboring rancher would be only too willing to lease that land next year, perhaps forgetting that he, too, would be trapped into buying a bottle of whiskey, and that he, too, would have to tolerate coming upon what was left of one of his steers butchered by the same "redskins."

I fell in love that summer at a Saturday night dance. She was a beautiful white girl, and her magic was so overwhelming that I swear I failed to sleep the entire night after I met her. I never even held her hand except while dancing. A surge in maturity arrived in other ways that summer. As a teenager in North Topeka, I would have been denied access to a bar, certainly by family members. But in South Dakota, as a teenager, I learned more, and at lower tuition, than anytime before or since. For it was there that I scrutinized, with the civilized eye of a Kansas River Valley Methodist, drunk cowboys, married or not, who hugged and smooched young natives and from time to time disappeared into the shadows of the dusty back streets of White River.

The landscape was mostly unplowed then and still is today. The horse, central to that way of life then, is less so now. Out on the ranch, besides the moon and stars, the only lights were from the towns of Murdo and Okaton across the river twelve to fifteen miles distant. It was a summer of branding, castrating, and fence mending, of dens of rattlesnakes to discover and fear, and pond bass to catch. Many evenings on the ranch I'd drive out on "the point" in a Cadillac coupe or the pickup to shoot prairie dogs or to see the hundred head of Andrew's horses out on the range or in the bottoms. "Junk horses," Ina called them, for in the dry 1930s she'd pumped water for hours for the cattle, only to have Andrew's horses show up, run the cattle away, and drink all the water. Andrew justified keeping these mostly wild creatures around in the belief that it was horse trading that had made it possible for him to be so solidly positioned. Today, I am amazed to think of the slack Andrew and Ina enjoyed to be able to afford those mostly unbroken horses.

I lived in a small wooden hillside shack set up on steel wheels, a shack

Andrew had bought from Mellette County, which had used it to house the county road crew. Horses had pulled the shack to its present location over the county roads, perhaps the same horses that formerly pulled the grader blade. Forty yards away Andrew and Ina lived in a small two-room house with a large attic. The ceiling bowed from the weight of old issues of magazines such as *Life*, *The Saturday Evening Post*, and Ina's favorite, *True Stories*, a magazine devoted to idealized romance. There was no electricity, only cistern water that was used at least twice, the last time always to water either a small backyard garden or the chickens. Some evenings Andrew and I would sit on the entry steps to their house, which overlooked the White River a half mile away, and Andrew would cuss Roosevelt, cuss the Yalta Conference, cuss Indians and neighbors and everybody but Ike the President—who happened to be Ina's first cousin, or I suspect he would have caught it, too.

But it was Ina and I who rode the range together. With Ina on her buckskin, Dickey, and me on Bonnie or Violet (names I picked from two of my girlfriends back home), we rode the range from one dam to another, where we kept poles with lures so that we might catch some bass. Or in the pickup, which I always drove, we might go to the abandoned school on the "school section" for some cottonseed cake to distribute to cattle somewhere across the nearly four thousand acres of paradise. When I drove the pickup or Cadillac coupe too fast, Ina referred to me as a "prune juicy boy," relentlessly ordering me to slow down. Toward summer, the cattle were rounded up to be trucked to the railyards and shipped by rail to Sioux City. The three of us drove the Cadillac to Sioux City and the stockyards. From there I rode the train to Kansas City, where I met the bus to Topeka. I did not want to go home, and had it not been for high school football in September, I might have stayed. The place became my American dream.

And yet, looking back, even though Lewis and Clark's Missouri River was only fifty miles downstream, I now see that little of Thomas Jefferson's vision of the yeoman farmer was ever possible there. On those prairies the land determined what could be done if humans were to stay. Some tried to farm the upland flats, but mostly failed. I loved everything about that country—the Indians; the rodeos; the Danish and Swedish immigrants, some with heavy accents, but all delighted with their landholdings; the rattlesnakes (from a distance); even the colorful prejudice. I loved the way the natives got a little bit even—with the butchered steer, the grocery bill, and the whiskey.

In the Kansas River Valley it had been another story. We were farmers there. Hoeing was endless during the summer, what with watermelons, sweet potatoes, cantaloupes, strawberries, peonies, phlox, sweet corn, potatoes, tomatoes, rhubarb, asparagus, and more. So it was a relief to put up alfalfa hay or harvest wheat, rye, and corn (my dad won the county corn-growing contest at least three years). We had a market along Highways 24 and 40, a two-lane highway when our county had between half and one-third as many people as it does now. People were heading west, and so it was often called the Pacific Highway. In our area it followed the path of the Oregon Trail.

Six children were born to my parents. The first, a daughter, was born in 1914; I was the last, in 1936. Dad was fifty that year; my mother, forty-two. It took me decades to realize they were Jeffersonian agrarians—and fiercely so, I see now. They were also Methodists and Congregationalists: Don't waste time, motion, or steps. Don't drink pop or alcohol in any form, or eat out. The contrast between that truck, grain, and hay farm and the South Dakota ranch was striking. My parents' row crops required cultivating and hoeing. Sweat of the brow, good manners, and quoted scripture

Figure 1. Home of Andrew and Ina Swan on their ranch north of White River, South Dakota, in Mellette County (photo courtesy of Luree Wacek)

went together. And in the market, our family met and talked with numerous people from out of town who were stopping on their way, often from coast to coast. America was on the move, mostly westward it seemed. Perhaps it was my age, but it seemed to me that we were countrymen then in a way that we are not now. No bad jokes about either California or New Jersey then. We all inquired into one another's well-being.

But back to our farm. Its main feature was the row crops. I will say that I loved much of that farming. But it did not compare to the life of the range, with natives and grassland, ranchers and rodeos. I made up my mind that I would have that South Dakota ranch one day, or one like it.

Figure 2. Andrew and Ina. Date and photographer unknown. Probably mid to late 1940s (photo courtesy of Luree Wacek)

But Andrew died of prostate cancer, and Ina died of injuries sustained in a wreck with her pickup. The ranch was sold, and the money went mostly to Andrew and Ina's nephews.

Football and love kept me in college in what must have been one of the most misspent youths in history. What smoldered in me were two experiences with land: the Jeffersonian agrarian, where culture dictated that ground be plowed and worked, and the life of the cattleman. I preferred the latter.

On May 30, 1854, the Kansas-Nebraska Act was ratified, creating the two new territories and allowing settlers to determine if they would allow slavery within their boundaries. Only two weeks earlier one of my great-grandfathers, born in England, entered Kansas from Pennsylvania. Twenty-six years old, he had already been to San Francisco and back by way of Panama. Fifty miles into Kansas, he broke tallgrass prairie sod and set right to farming, Jefferson-style. Two years after arrival, he interrupted his home-steading efforts to fight alongside the abolitionist John Brown in the first bat-tle between free state and proslavery forces at Black Jack on the Santa Fe Trail. He later bought the land where that battle occurred and planted sugar maple trees from which sugar was once harvested, trees that stand to this day.

The man who was to become his son-in-law, my grandfather, arrived in Kansas in 1877, one day before turning twenty-two, with $300 to his name. He felt lucky not to have put his money in the bank, for it closed the next day. He thus preserved his grubstake and threw himself out onto the Flint Hills grasslands of Kansas to run cattle on more or less free grass. At the end of ten years he had enough to go in with a partner and purchase 160 acres of sandy loam in the Kansas River Valley on the second bench above the riverbank, thereby assured of no more than a flood or two per century. After a few years he bought his partner out.

I was born on that farm at the height of the Great Depression. I still love those soils, love to plow them, love to smell them. As good as it is, I have wondered why that grandfather who had purchased the farm, when the grass had been so good to him for cattle, would give up that way of life to farm. It has taken me decades to acknowledge the power of culture and regional history, the power of a worldview. Before Kansas he had lived in the Shenandoah Valley of Virginia. A Virginian! He was an agrarian! As Aldo Leopold has reminded us, nothing so important as an ethic is ever written. Rather it evolves in the mind of a thinking community. In Leopold's view Moses didn't write the Decalogue, but rather summarized it for a seminar. In a similar manner, the ideal of the family farm as the source of virtue for the yeoman farmer had a history long before Jefferson articulated it. Nevertheless, I could not resist, when thinking about the agrarian ideal, looking at my road atlas, finding the scale and using my ruler, and noting that my grandfather grew up less than fifty miles as the crow flies from Jefferson's Monticello. Because of the history of that worldview going back at least to the Greeks and Hebrews, the meaning of that distance in space is far less significant than the distance in time. Jefferson's ideal was a cultural handing down, an ideal featuring a small family farm where pleasure and work go together. This wasn't learned from the formal culture of textbooks.

Farming paid handsomely to farmers in my grandfather's time and allowed him and his family to play the Jeffersonian role. They played it well, for he and his family were comfortably well-off when he died in 1925. He was not alone. To illustrate: it seems difficult to imagine in our time that he and his neighbors could have enough surplus income to decide that the new school should be completely paid for in the year it was built. It was a fine, well-built school, two rooms for eight grades. My mother and I both went there. The community could pay such a debt so quickly, I suspect, because most of the members of that community had a shared cultural understanding, that they were not inclined to expand their income by expanding their acreage. Profits from farming at that time and place were sufficient, with some sacrifice, to pay it off. Plenty of good farmers were around then, people who enjoyed farming. I am remembering an offhand statement my mother made once about her father. She said that he would lean on his scoop shovel or against the barn for a half hour or more, watching his hogs eat the ear corn or the soaked oats or the boiled potatoes he had raised. As much as his experience was a reflection of the times, it was also a combination of joy, sympathy, art, and love rolled into one and tuned to the demands of his

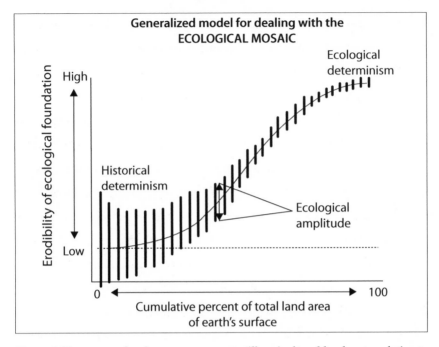

Figure 3. The area under the curve represents till agricultural land accumulating to 100%. Most agricultural land is erodable. At the extreme left, where the vertical bars are long, the degree of forgiveness following disruption of the original vegetative structure is illustrated. This could be river valleys such as the Kansas River Valley or the Nile of Egypt. A variety of crops can be grown and the degree of forgiveness of human error is large. The short vertical lines in the upper right may represnt a steep south-facing slope in the intermountain west of the U.S. or the ranchlands of South Dakota described in the text. Here one had better have a vegetative structure similar to what was originally there or there will be a loss of ecological capital. The degree of forgiveness is small.

place. Here was the Jeffersonian dream, as imperfect as it was, at its high-water mark. The actuality or reality of that dream has been compromised and in decline ever since, and since World War II at dazzling speed. In fact, now it seems as if the Jeffersonian dream, as Aldo Leopold once said about conservation, is "a bird which flies faster than the shot we aim at it."

It was only much later that I had the language to understand and describe how on the South Dakota ranch, ecological determinism is the major factor—how the land, not humans, determined behavior. On those South Dakota prairies, west of the Missouri, if the original vegetative structure is plowed, a steep reduction in carrying capacity will shortly follow. Specifically, the flow of nutrients down the White River, downward toward the river Jefferson's emissaries traveled—the mighty Missouri—amounts to the loss of ecological capital necessary to sustain a diversity of organisms. In the Kansas River Valley, however, historical determinism rules the soils, and Jefferson's ideal can be a reality. I am saying that because of its soil and vegetative structure, South Dakota requires grazing. The Jeffersonian ideal can exist larger in Kansas because of soils and rainfall. It is a more forgiving landscape. Across the ecological mosaic exists a range between and beyond these two extremes. It is time for each piece of that mosaic to be evaluated according to where it falls on that scale.

DISCOVERING ALDO LEOPOLD

I am embarrassed to say that a full decade passed from that time on the South Dakota ranch before I would become aware of Aldo Leopold and begin to adopt his understanding. It was 1963, and I was twenty-seven years old, and it would be another decade before I would fully internalize the implications of his insights for a new agriculture. Since the start-up of The Land Institute, more times than I can count, I have had to handle with forced composure the question, "Who is this man Aldo Leopold you keep talking about?" And over the years I've tried various answers: author of *A Sand County Almanac*, preeminent leader in the conservation movement, composer of a brilliant statement he called "The Land Ethic."

My answers have always sounded feeble, partly because Leopold's thought was so comprehensive, partly because Leopold's life was so comprehensive. A typical follow-up question has been, "What does he have to do with sustainable agriculture?" Again, my answers have seemed incomplete, due in part, I suspect, to the completeness of the man.

In talking about Leopold's relevance to agriculture, I have often used a term now enjoying increased usage: *robust*. Not in the sense of a robust athlete, but in the more formal sense of something such as a law that holds true across a wide range of observed phenomena, bridging such disciplines as ecology, economics, and physics.

Leopold's insights were "robust" because he kept himself abreast both of older principles and of emerging developments in ecology. He was also a student of history, society, and culture. Such a field naturalist is more than a student of predator-prey or host-parasite relationships, more than a student of the food chain. His was an integrated life, one in which the insights of his expansive scholarship made him a prominent member of a select ecological academy. Without labeling it, he recognized the *problem of agriculture*. He saw a reality inherent in all of nature's ecosystems, a reality at work since the earliest days of life on this planet: Nature's ecosystems feature material recycling (meaning the exchange of materials between living and nonliving parts with minimal loss seaward from the farm) and run on contemporary sunlight (as opposed to fossil fuel). Most human ecosystems do not.

Through this special lens he examined our history and fate as tillers of the soil. Leopold understood that agriculture is a recent phenomenon and a departure from the way nature has worked through hundreds of millions of years of evolutionary change. He also understood that when we disrupt the diverse integration of species (as we do with farming), the ecosystem will decline—which means that the harvest of contemporary sunlight also will decline, for without the soil-sponsored nutrients, land plants are unable to capture the carbon, hydrogen, oxygen, and nitrogen bound together in

Figure 4. Aldo Leopold and his shack (photo courtesy of Leopold Center)

water and carbon dioxide afloat in the atmospheric commons, and store the sun's energy through photosynthesis. The attempts by early agriculturalists to meet specific dietary requirements on a broad scale expanded the Neolithic garden into what we call fields. And we continued to ratchet up the scale of fields even as we expanded into more of nature's lands, disrupting the natural integrities and depleting fossil carbon.

Beyond Leopold's ecological awareness was his deep sense that the human, as a sociopolitical creature, is either ignorant of or cares little about the discontinuity between nature's way and the way of the agricultural human. Leopold's frustration on this point is clear. His essay "The Land Ethic" is at once a philosophical discourse and a prototype for a moral code. Anticipating the changes still to come, he wrote, "We can bolster poundage from depleted soils by pouring on imported fertility, but we are not necessarily bolstering food-value."

Another one of my favorite essays is "Odyssey." Here Leopold follows an atom called "X" through the world of the Native American. He then follows another atom called "Y" through the world of the immigrant settler-farmer. The greatly accelerated speed of "Y" tells the story of land exhaustion through agriculture. The poor attempt on the part of the civilized human to rectify this problem is expressed in one sentence: "So the engineers built pools like gigantic beaver ponds, and [atom] Y landed in one of these, his trip from rock to river completed in one short century."

Perhaps the most important essay relevant to my education after "The Land Ethic" is "Thinking Like a Mountain." In it he describes watching the "green fire die in the eye of the old wolf." As moving as it is, so also is the comment that follows—that the mountain knew better than to have the wolves killed in the interest of having more deer. In one bold statement, the mountain ecosystem is a slab of space/time whose health can be measured when its integrity has been disrupted by the killing of a predator.

I could stop here were not the opportunity so rich to go on and explain how Leopold's ecological savvy turned him into a first-rate philosopher. His ecological insights caused him to challenge basic assumptions of modern science. More than any notable ecologist since Charles Darwin, Leopold's mind ran against the tide of Baconian-Cartesian thinking.

In 1887, when Leopold was born, Darwin's ideas were not three decades old. Leopold died in 1948, eighty-nine years after the publication of *On the Origin of Species*. In that short time no one in the conservation movement had internalized the Darwinian evolutionary ecological worldview more completely than Aldo Leopold. Few, if any, saw ecology as the discipline that would challenge the worldview of Bacon and Descartes. Of course, agriculture had risen long before Francis Bacon and René Descartes, but their scientific worldview drove (and still drives) the modern trend toward the industrialization of agriculture. Leopold provided much of the intellectual framework toward the eventual marriage of ecology and agriculture.

"The Land Ethic," "Odyssey," and "Thinking Like a Mountain" made it possible for me to "see" the South Dakota ranch, its cultural and historical underpinnings, in contrast to the near-opposite farm where I grew up.

BEN W. SMITH'S THIRTEEN-WORD SENTENCE

Another step in my education came in 1965 in my lab in the Genetics Department at North Carolina State University. I was sitting behind my microscope one evening looking at chromosomes when my professor, Ben W. Smith, stepped in the doorway, holding his pipe, and delivered a single sentence: "We need wilderness as a standard against which to judge our agricultural practices." He left before I could reply.

I had a thirteen-word sentence that would lead me to a still-incomplete discovery.

More than ten years later, in 1977, I led a group of Land Institute students on a field trip to the Konza Prairie, where the late, great Kansas State University ecologist Lloyd Hulbert was conducting a tour especially for us. I had just read a General Accounting Office (GAO) report that pointed out what appeared to me to be a sorrowful reality: soil erosion was as severe as it had been three decades earlier during the dust bowl era of the 1930s, when the Soil Conservation Act was formed. Over the next several hours I listened to Professor Hulbert describing the root dynamics of the Konza Prairie ecosystem and the various roles played by members of the biotic web.

That GAO report could have been just another document among my countless unorganized piles saved to be better read when I had the time.

But after the trip to the Konza Prairie and with my mind on the report, I drew a branching tree diagram and scratched in sixteen potential combinations involving four paired contrasts: polyculture versus monoculture, perennial versus annual, woody versus herbaceous, and fruit/seed versus vegetative. Among the sixteen combinations, four are nonsensical (for example, woody annuals don't exist). Of the twelve possibilities, eleven of the blanks had been filled, meaning they were useful for direct human purposes. But one combination, a polyculture of herbaceous perennials devoted to fruit/seed, was blank. Why this was so I could not get off my mind. It became clear that our failure to fill that blank stood behind the ten-thousand-year-old "problem of agriculture." So in 1977 I wrote a paper entitled "The Search for a Sustainable Agriculture," which was published in the Friends of the Earth publication *Not Man Apart*. In 1980, Friends of the Earth published my book *New Roots for Agriculture*, an expansion of the idea, and supporting argument of its necessity and possibility.

Poly vs. Monoculture	Woody vs. Herbaceous	Annual vs. Perennial	Fruit/Seed vs. Vegetative	Current Status
1. Polyculture	Woody	Perennial	Fruit/Seed	Mixed orchard (both nut & fleshy fruits)
2. Polyculture	Woody	Perennial	Vegetative	Mixed woodlot
3. Polyculture	Herbaceous	Annual	Fruit/Seed	Mixed cropping (corn-beans in the tropics)
4. Polyculture	Herbaceous	Annual	Vegetative	Dump heap garden,* companion planting
5. Polyculture	Herbaceous	Perennial	Fruit/Seed	
6. Polyculture	Herbaceous	Perennial	Vegetative	Pasture & hay (native or domestic)
7. Monoculture	Woody	Perennial	Fruit/Seed	Orchard (both nut & fleshy fruits)
8. Monoculture	Woody	Perennial	Vegetative	Managed forest or woodlot
9. Monoculture	Herbaceous	Annual	Fruit/Seed	High-producing agriculture (wheat, corn, rice)
10. Monoculture	Herbaceous	Annual	Vegetative	Ensilage for livestock
11. Monoculture	Herbaceous	Perennial	Fruit/Seed	Seed crops for category 12
12. Monoculture	Herbaceous	Perennial	Vegetative	Hay crops (legumes & grasses) & grazing

*See *Plants, Man & Life* by Edgar Anderson for the splendid chapter on dump heap agriculture

Table 1. Twelve possible combinations with one blank

34

WENDELL BERRY

With that book began my personal connection to Wendell Berry (see color insert *fig. 3*). It started with a letter from him dated November 11, 1980. That opened with a compliment of my book. But rather than wait for me to let the praise soak in, in the next paragraph Wendell pleaded for me to reconsider the role of draft farm animals in the utopian future I had outlined in my last chapter.

He concludes his letter with the following paragraph:

> As one who has farmed with both tractors and teams, I would insist (to you; I would be more cautious, at present, in a public statement) that with the use of a tractor certain vital excitements, pleasures, and sensitivities are lost. How much numb metal can we put between ourselves and our land and still know where we are and what we are doing? Working with a tractor is damned dulling and boring. It is like making love in boxing gloves.

I answered his letter. I took his critique seriously. Our next connection was over the phone one evening. He said, "This is Wendell Berry." I responded, "Oh, my gosh," then turned and announced to the whole family, "It's Wendell Berry." My admiration for his writings, especially *The Unsettling of America*, had already placed him at the top of my list of modern writers. With *The Unsettling* I had found an ally. Long before I finished reading that very great book, I felt less lonely.

Wendell placed the call to say that he wanted to do a story on The Land Institute for Rodale's *New Farm*. The Land Institute is a nonprofit educational research organization in Salina, Kansas, devoted to the study of sustainable alternatives in agriculture. Wendell came for a visit. We toured the premises, looked at the various projects, talked of our families, of our origins: I had grown up in the Kansas River Valley; his home looks over the

Kentucky River Valley. It was as though we had the brands of our rivers and states on our backs. We talked about plans we had in mind for our places.

Straightaway Wendell expanded my education as he introduced me to my intellectual ancestors, people I never knew but who had informed the agrarian culture, the cultural handing down of the likes of Liberty Hyde Bailey and Sir Albert Howard. I had known of Liberty Hyde Bailey through his manual of cultivated plants, but not Bailey the cultural agrarian, and not Albert Howard at all, though I had seen references to his ideas. We both had read *Tree Crops* by J. Russell Smith, another agrarian.

Here was this Kentucky farmer, educated in the humanities, who had gone home. Trained in the sciences, I had gone home, too. The fact that we had come from different disciplines was of no consequence. We came together in what we both liked—land and farming. Wendell's side trip away from his origins into the professional world of literature must have helped him adjust his lens and see his place for what it was. My side trip into genetics was sponsored in no small part by our Kansas River Valley farm, where heredity had been my earliest interest. My curiosity about family and neighbor likenesses and gestures, as well as livestock traits, made genetics an easy discipline for me, and it was also the closest I could get to the culture of agriculture in academic life. But I couldn't get close enough. At some level I knew that what I loved about farm life was disappearing faster than I could apprehend it through my academic interests.

So I came home and found my life's work. But I was a late arrival. Wendell had come home several years before, and once home found his subject. Here I must pause, so as not to be misunderstood. Wendell used his subject in a way different from that of most artists. Rather than regard his homeland, his people, his community, as a mere reservoir for his art, he assumed the responsibility of defending that land, defending those neighbors, defending that community, and by doing so defending rural culture everywhere with his art. It is his eloquent defense that has heartened all of us who had some inkling of the need for connectedness that he speaks about so richly. He has taught us its necessity and possibility. And while few of us even yet can be optimistic, Wendell has given us enough to be hopeful. His novels, such as *Watch with Me*, *A Place on Earth*, and *Jayber Crow*, accomplish this alone. His essays accomplish this alone. His poetry certainly speaks to all of that and more. I can say without hesitation or qualification that Wendell Berry may be the most practical person I know, precisely because he made his art capture the roundness of a rooted, grounded life—beyond acres and calories, foundations and roofs, crops and livestock.

It wasn't long before Wendell introduced me to some of his friends, his fellow agrarians—Maurice Telleen, publisher and editor of the *Draft Horse Journal* (Maury, the man Wendell dedicated *The Unsettling of America* to), and then farmer and writer Gene Logsdon. Among those three there was a trinity, unholy in the conventional understanding but whole, in fact the whole I needed. In July 1983 he introduced me to David and Elsie Kline and their Amish family in Ohio. These were real allies. The conversation among all of us continues by letter and by phone. Wendell gave us the language of a necessary culture coherent with our ideas about an ecosystem-based agriculture. Here is a sample.

On April 20, 1981, came a letter from Wendell with echoes of *The Unsettling*, in which he speaks to the declining number of defenders of the practical necessity of small-scale farming:

> Who is going to defend [the small farm]? If its justification were only cultural or political, then one might cynically give it up. But I am convinced that it is necessary to the survival of agriculture

too, that, in general, the best husbanded, most productive farms are small. I believe I have looked carefully at as many farms as *anybody*, and I'm convinced that the physical, practical evidence is overwhelmingly in favor of small-scale, balanced structure, and diversity—along with the necessary cultural supports.

The cultural element described by Wendell was totally consistent with the ecosystem concept we were exploring at The Land Institute. Wendell has relentlessly featured people, land, and community as one. No mere nostalgia here; art and the practical necessity are one. A few months later he went on to describe a world that is compelling to so many, and showed the possibility of its being widespread:

August 22, 1981

I was standing at the corner of Ed Poe's little cattle shed the day before yesterday (Ed is the neighbor who was mowing grass when we talked to him). All cornering in there within easy range of a cheap camera were a small pasture, a one-acre tobacco patch, a one-acre field of soybeans (for hay), and a patch of potatoes (for eating), patches of woodland on two sides and not far away. And men of four households at work in the shed, putting up a harvest of about 400 bales of excellent bean and millet hay from about two acres of very "marginal" ground. Pleasant work, good economy, truth, and beauty. One big tractor would ruin it all.

Here was the agroecosystem with a human community embedded. Working against those men of four households was a combination of economics made possible by low-priced nonrenewable energy. It was perhaps this paragraph that caused me to consider the possibility of a general rule: high energy destroys information, in this case the information of both the cultural and biological varieties. On Ed Poe's farm we see the human scale, the "small is beautiful" reality described by E. E. Schumacher in the 1970s.

Here was the "culture" of agriculture. Wendell described the cultural information in the workers and the biological information stored in the genes of the crop diversity.

My education with Wendell continued. That same year I had several conversations about the domestic and the wild. I had argued that sustainability would rest on a three-legged stool. One leg was conservation of the wild biodiversity, another leg was restoration, and the third leg was a well-built agriculture or forestry or fishery such that, to use the words of historian Angus Wright, "conservation was a consequence of production." Some ecologists argue that to save the biodiversity we need to intensify the industrial processes for food production "where it is already screwed up." Here is Wendell's response:

May 30, 2002

The job we have now is to oppose the proposition that natural diversity and the integrity of the natural world can be preserved (1) by making a strict division between the natural world and the human world and (2) by radically reducing the cultural, economic, and domestic-genetic diversity of the human world. . . .

Can we preserve nature intact in half the world by utterly degrading it in the other half?

Can we preserve nature intact anywhere by radically oversimplifying our economic relation to it, dismissing nature's integrity as an economic standard, and ignoring its processes—in effect, replacing it with the standards and processes of industrialism?

Can we afford to abandon forever the possibility of a living harmony between humanity and nature, economy and ecology? It is just to say "forever" here, because if we destroy all the land-using or land-based cultures, we may be unable to develop them again—not, at least, with the present population in the foreseeable future. . . .

What are the political implications of reserving half the world for nature and delivering the other half to the corporate economy?

What governmental, police, and military powers and measures will be required for that? There is no possibility here of separating science from politics and industry. This, in fact, looks like a deliberate relinquishment of science (and every other discipline) to politics and industry....

Finally, what is to keep the corporations, once they have demolished the natural integrity and diversity of the human half of the world, by the prescribed industrial methods of land use, from going directly on to apply the same methods to the natural half? (I'm adapting my language only by courtesy to the absurd notion that nature can be whole in half the world.)

So, as I see it, the long-building opposition between agrarian conservationists and puritan conservationists is now becoming public. That's too bad. In many ways it will be destructive. But I see no alternative to standing up and defending the side we're now forced to take.

As I see it, Wendell has authored the modern challenge for all of us interested in conservation, which brings us to another major contribution of his. It involves one of the most useful intellectual exercises, at once literary and historical. It came to my attention in the 1980s at the dedication of our research greenhouse at The Land Institute. Wendell was the featured speaker, and it was here that he spoke to the long history of "nature as the measure" in both the literary and the scientific tradition.

Various friends of The Land were crowded into the new headhouse of that facility. Settled in, we all sat back as Wendell traced the notion of "nature as measure," describing its roots at least two thousand years before Jesus of Nazareth. He began by quoting Job: "Ask now the beasts and they shall teach thee, and the fowls of the air, and they shall tell thee: Or speak to the earth, and it shall teach thee; and the fishes of the sea shall declare unto thee." He quoted Virgil who, at the beginning of *The Georgics* (36–29 BC), instructs us that "before we plow an unfamiliar patch / It is well to be informed about the winds, About the variations in the sky, / The native traits and habits of the place, / What each locale permits, and what denies." Toward the end of the sixteenth century, Edmund Spenser called nature "the equal mother" of all creatures, who "knittest each to each, as brother unto brother." Spenser also saw nature as the instructor of creatures and the ultimate earthly judge of their behavior. Shakespeare, in *As You Like It*, has the forest in the role of teacher and judge. Milton, in *Comus*, has the Lady say of nature, "she, good cateress, / Means her provision only to the good / That live according to her sober laws / And holy dictate of spare Temperence." Finally, Wendell drew on Alexander Pope, who in his *Epistle to Burlington* counseled gardeners to "let Nature never be forgot" and to "Consult the Genius of the Place in all."

After Pope, Wendell noted that this theme of a practical harmony between humans and nature departs from English poetry. Later poets saw nature and humanity radically divided. A practical harmony between land and people was not on their agenda. The Romantic poets made the human mind so central that nature became less a reality to be dealt with in a practical way than merely a "reservoir of symbols."

One wonders what the consequences would have been if the settlers and children of settlers whose plowing of the Great Plains in the 1910s and '20s gave us the dust bowl of the '30s had heeded Virgil's admonition that "before we plow an unfamiliar patch it is well to be informed about the winds." What if they had heeded Milton's insight about the good cateress who "means her provision only to the good / That live according to her sober laws / And holy dictate of spare Temperence"? For both, nature gave the measure, the standard, the lesson.

What about a *science* of agricultural sustainability with nature as the measure? Scientists are not expected to speak in terms that echo the poets.

But a few have. Wendell noted that "nature as the measure" went underground among the poets in the nineteenth century. When it finally surfaced it was among the agricultural writers who had a scientific bent. Liberty Hyde Bailey's *The Outlook to Nature* appeared in 1905. That grand old Cornell dean described nature as "the norm": "If nature is the norm then the necessity for correcting and amending abuses of civilization becomes baldly apparent by very contrast." Later, in *The Holy Earth* (1915), Bailey advanced the notion that "a good part of agriculture is to learn how to adapt one's work to nature. . . . To live in right relation with his natural conditions is one of the first lessons that a wise farmer or any other wise man learns."

Sir Albert Howard published *An Agricultural Testament* in 1940. Howard thought we should farm as the forest does, for nature constitutes the "supreme farmers": "The main characteristic of Nature's farming can therefore be summed up in a few words. Mother earth never attempts to farm without live stock, she always raises mixed crops; great pains are taken to preserve the soil and to prevent erosion; the mixed vegetable and animal wastes are converted into humus; there is no waste; the processes of growth and the processes of decay balance one another; ample provision is made to maintain large reserves of fertility; the greatest care is taken to store the rainfall; both plants and animals are left to protect themselves against disease."

Another of Wendell's conclusions is profound. He observed that among the poets and scientists he quoted—from the agrarian common culture—there is a *succession* in thought, but in the *formal* culture— meaning the teachers and students—the thoughts arrive now and then and therefore form a series rather than a succession. It is not surprising that the agrarian common culture provided the succession. But why the formal culture never established and maintained a succession on the

Figure 5. Liberty Hyde Bailey at Cornell University, 1925

writings of those poets and scientists needs to be answered. After all, both came out of the common culture.

But now is our moment. We have both the necessity and the possibility to build that succession in the formal culture, in the U.S. Department of Agriculture, in our colleges of agriculture. To do so, we have to quit trying to understand agriculture in its own terms and build on the science of ecology and evolutionary biology. We need more people who will show us the *practical* possibility of a research agenda based on a marriage of agriculture and ecology. That agenda will require a push from those who, after examining the assumptions of modern agriculture versus what nature has to offer, decide in favor of learning from nature's wisdom.

Learning from nature's wisdom means that we look to natural ecosystems precisely because they have featured recycling of essentially all materials and have run on contemporary sunlight. I say "feature" because they have not been perfect in those recycling efforts. (Not all life-forms are powered by the sun. The examples, however, are not worth mentioning in the context of the subject of this book.) Ecological standards based on studies of ecosystems that have experienced minimal human impact provide us with our best understanding of how the world worked during the millions of years before humans arrived.

Wendell helped me see that the cultural reality of agriculture is necessary to the ecological reality of agriculture. This is at once *possible* and necessary. The dualism between the world of wild biodiversity and agriculture does not have to be. There does not have to be a sacred and a profane on our earth. Wendell would not allow such thinking. He has worked relentlessly to stop it.

TWO ECOLOGISTS AND A CONCEPTUAL TOOL

I have had other mentors. Most notable are two ecologists. J. Stan Rowe served for many years as an ecologist in the Department of Plant Ecology and Crop Science at the University of Saskatchewan. Arnold Schultz, also an ecologist, spent his entire career as a faculty member in forestry at the University of California at Berkeley. Their work and thought round out the broad outline of my still-evolving synthesis. The two men overlapped as graduate students in ecology at the University of Nebraska under the famous ecologist J. E. Weaver, but had little interaction with one another there. Both were field ecologists. Both had a keen knowledge of the history of science in addition to their understanding of the history of ecology. And perhaps even more telling about the quality of their intellect is the fact that, possibly coincidentally, both changed their minds over the same three- or four-decade period about what an ecosystem is. Who knows what was at work? It is probably a combination of their ability and that era in the history of ecology. Whatever it was, both concluded that the ecosystem is not a mere container, but rather is a whole entity with properties of its own. I got to know both of them through correspondence and personal interaction on field trips. They allowed me to question them closely about their ideas and how those ideas relate to our work at The Land Institute. Their inspired insights have contributed so much to my journey.

Arnold Schultz

Arthur Tansley used the term *ecosystem* in 1935 (Roy Clapham had coined the term in 1930). Tansley admitted his biological bias, but was nevertheless a broad thinker in going beyond the biota. He argued that since *community* does not include soil, rocks, air, and water, and because *ecosystem* is more inclusive, it is a more logical unit for study. In 1942 the ecologist Raymond Lindeman gave this formal definition of *ecosystem*: "The ecosystem (is) the system composed of physical-chemical-biological processes active within a space-time unit of any magnitude."

Arnold Schultz arrived at the University of California–Berkeley in 1949 and discovered to his surprise that "no ecology grad students were studying ecosystems, no resource managers were thinking in terms of ecosystems, or anybody else until I met the one professor who had been thinking 'ecosystems' for a long time, Hans Jenny." The late Hans Jenny is probably the most revered soil scientist of our time. Jenny had introduced the expression "the larger system" in his book *Factors of Soil Formation*, which was published in 1941. The first to present the ecosystem concept in textbook form was E. P. Odum, who was teaching at the University of Georgia, in his *Fundamentals of Ecology* (1951).

Some time later Arnold was thinking about better ways to teach ecology. He had been doing quite a bit of reading about the systems approach, especially systems theory. "But in talking to my ecologist colleagues on the Berkeley campus," he recalled, "I found that this systems stuff was like a foreign language."

Arnold was involved in a collaboration with Hans Jenny, studying the marine terraces in Mendocino County, California. Arnold had what he would call an awakening during a conversation with Jenny. Jenny admitted, "About ecology I know very little, but about ecosystems I know a lot." Arnold remembers that "when I thought about that statement I realized more than ever that the study of ecosystems was quite a different field than the study of ecology—*and it was not a branch of ecology.*"

In 1960, with no intention of trying to create a new field, Arnold Schultz invented the word *ecosystemology*. It is not a pleasant word to say or hear. That reality alone explains why Arnold was reluctant to use it. At this stage in his thinking, at least, he was simply considering the title of a course he was planning to teach in the forestry department at UC Berkeley. *Ecosystemology* is what he came up with.

In 1962, the year that Rachel Carson brought out *Silent Spring*, Arnold submitted a paper to a faculty committee titled "The Ecosystem as a Conceptual Tool in the Management of Natural Resources." Five years later it appeared as a chapter in a book entitled *Natural Resources: Quality and Quantity*, edited by S. V. Ciriacy Wantrup and James J. Parsons. (I didn't see the paper until 1978, and then because my friend, UC Berkeley professor Dan Luten, a friend of Schultz's, called it to my attention.)

The committee members liked his approach, and so he developed an interdisciplinary graduate-level seminar under the title "Natural Resource Ecosystems." He rejected calling it "Ecosystems Analysis" because he felt it should have more synthesis than analysis. The seminar must have had some intriguing content, for when it was first offered in 1964, forty graduate students and twenty auditors attended. These sixty people represented thirty-two majors. In the second year eighty students enrolled; year three, doctoral and master's candidates from forty-four University of California–Berkeley majors attended (as well as students from Stanford, UC Davis, and San Francisco State). Over twenty-seven years, Arnold was relentless in stressing "holistic, interdisciplinary, and systems thinking." It spawned two new fields on the campus: agroforestry and landscape ecology in 1982 and 1987, respectively.

Then came the difficult, if not turbulent, period for Arnold Schultz at Berkeley, beginning around 1976 and continuing for more than twenty years. He had detractors of intellectual high standing on campus and elsewhere. An author of one of the largest books on plants was harsh in his criticism: "People who study ecosystems are dumb; they don't know anything."

Nevertheless, other ecologists besides E. P. Odum included the idea in textbooks. The International Biological Program spent millions of dollars doing ten-year studies on ecosystems. But even so, as Schultz noted, one well-known ecologist tried to quash their value by writing, "Now that the ecosystem fad has finally been put to rest, we can get on with studying ecology." (Schultz said this statement is paraphrased a bit since he could not find the exact quotation.) He remembers a UC Berkeley ecologist saying, "It's all in your head, and pretty well muddled. There's no such thing as an ecosystem." This comment foreshadowed more trouble. Systems Ecology was taught but soon dropped as a core course. Schultz's arctic study didn't help California farmers, so what was the use? Many thought his pygmy forest study was interesting but not practical. No forestry graduate students were willing to use it in thesis research. Schultz and Hans Jenny applied for a research grant from the California Division of Forestry for studies in the

Jackson State Forest. Schultz's department chairman, one of the referees, saw no value in the proposal. Ecosystem study had sunk in value by 1980.

Finally there was a little traction. In the College of Natural Resources at UC Berkeley, a new division was established called Ecosystem Sciences, which includes such fields as soils, hydrology, and biochemistry. These old disciplines are now *integrated* but, as Schultz believes, should point to just *one* Ecosystem Science, not the plural form the title suggests. Schultz again: "A problem I have with many of the ecosystem studies that have been published is that the investigators use the ecosystem as a *container*. In other words, they may study processes, populations, or communities *within an ecosystem*, but not the ecosystem itself as a whole entity, with properties of its own. Most ecologists remain comfortable with the classical methodologies of population or community ecology." They keep it plural.

Understanding that distinction—"ecosystem as a container" versus "ecosystem as a whole entity"—has become increasingly important in our time, in that the tropical rain forest, for example, is more than a trivial factor in regulating the hydrological cycle of our planet. Landscape ecology gained credibility as a quasi-discipline during the dust bowl years in the Great Plains during the 1930s. It seems that now and then we get a grip on the necessity to expand the scale to the appropriate level, but what we might call "the problem of reassertion" rises up. For example, the plural "Ecosystem Sciences" provides license to investigators to keep knowledge in its more reductive categories, and therefore to experiment and resolve ambiguity only to relatively small questions. The singular form forces us to ask questions to which there may be no available answers. These very questions, however, could lead to "yeasty" thought, useful if we are to break the stranglehold of reductionism, in which priority is placed on the part over the whole. As Richard Levins and Richard Lewontin have suggested, to be reductive is not wrong. The error lies in believing the world is like the method.

The ecosystem concept, as understood by professors Schultz, Jenny, and Rowe, may have been slow to catch on simply because of a cultural lag within biology. We all know that ecology as a discipline was slow to be accepted before 1970. It is still the most pecked over of all the disciplines within biology. Arnold Schultz speculates that if the concept had been pushed by philosophy (as *wholeness*), by systems theory (through *modeling* and *design*), or by methodologies (from other fields) instead of only from ecology and management, it would have had broader appeal from the start. Arnold acknowledged the standard definition of ecology, but also asked what had happened to the term and concluded, "Not much, except that *relations* became the key word, *systems* was substituted for organisms, and environment got blown up out of all proportions!"

Stan Rowe

And then there's Stan Rowe. How is it that Stan Rowe could come up with a way of thinking that differed from most biologists? Certainly, his field experience, which included flights at low altitudes across Canada's vast spaces, was instrumental. He had numerous opportunities to look at "the larger system," and landscape ecology was part of his purview. But others had these opportunities, too.

Stan was the son of a Methodist minister, but so what? He refused military service during World War II and was imprisoned for a period, but was released to teach Japanese students in camps set up by the Canadians. I would not want to discount for a moment that this history may have had a subtle impact on his thinking about the way our ecosphere, our earth, is.

But this is only to say that Stan was no ordinary sort of man from early on. A more tangible explanation resides in his having steeped himself in the history of science, biology, and ecology, which together with his personal history, as it did with Arnold Schultz, made him realize that the ecosystem was not a mere "container," but rather a "whole entity."

Stan pondered why the textbooks did not reflect this way of seeing the world, why the research in ecology was so increasingly reductive. He was not alone in his inquiry. As he dug in he was helped by such early systems thinkers in the last century as Ludwig von Bertalanffy and J. H. Woodger, who published "Modern Theories of Development." In 1945 Alex B. Novikoff published "The Concept of Integrative Levels and Biology." In 1950 Bertalanffy published "An Outline of General System Theory." These three papers were cited in a 1954 paper by James K. Feibleman entitled "Theory of Integrative Levels." Feibleman described twelve laws that apply to the hierarchy of structure, beginning with atoms, then to molecules, and on the biological side on to cells, to tissues, to organs and organisms.

Stan Rowe, along with others, was asking: What comes next after organism? Species? Population? What? He considered what the other levels had in common and noted that it was contiguous volume. With this insight it was easy. Species don't have contiguous volume. Populations don't have contiguous volume. Ecosystems do. In 1961, he published his landmark paper, developing the volumetric criterion for "thinghood," which placed ecosystem right above organism and therefore able to be regarded as a whole entity with properties of its own. It must have been a leap in intuition if not thought. The beauty of this insight was that Feibleman's twelve laws held, making the ecosystem the appropriate additional slab of space/time.

Stan didn't stop there, not by a long shot. For years, he continued his reading in the history of science, biology, and ecology, knowing he would have to do so if he was to get at the root of the unbalanced development in biological thought and experimentation. He eventually concluded that ecology had not outgrown its lowly status as the fourth field of biology. He went through the history of biology, noting that biologists first studied *morphology*, the structure of organisms, how they are put together in a mechanical sort of way to make bone, muscle, organs, whatever. Next they studied *physiology*, whose main feature is process: blood circulation, hormone action, all the way into chemistry. Third came *taxonomy*, a pigeonholing process. The taxonomist's purpose is to catalog and file the biotic diversity, mostly in cabinets and drawers. Carolus Linnaeus, the father of modern taxonomy, has inspired biologists for the past three hundred years to bring order to the biological world.

Ecology, as the fourth field, Stan said, was still perceived as a discipline that plays around the edges of biology instead of as a more comprehensive discipline that integrates biology with all the earth sciences.

Since most biologists were still thinking of the ecosystem as a container, Stan continued to press. Part of the problem, he insisted, is the word *environment* itself. Rather than organisms, including humans, seeing ourselves as enclosed within a "miraculous skin," the environment is regarded as "out there." The old subject-object dualism reigns. We betray ourselves when we say we are going to do something "for the environment" or "for the ecology." We sound as though we intend to meet some need with a favor, like giving money to the poor. This is a formidable problem precisely because of our relationship to the earth, which for most of us means looking from the inside out. It is commonsense or natural to do so. To illustrate and correct this perception, Stan asked us to imagine ourselves becoming small enough to go inside a cell, so small we would need binoculars to look around. In that cell we are apt to declare that some objects are nonliving,

Figure 6. J. Stan Rowe with bearded student Mark Gernes at The Land Institute, 1986

such as crystals. Still on the inside, that which wiggles or shows motion of some sort we, with our conventional minds, would regard as "alive." When we step outside the cell we have no trouble saying the whole cell is alive. And so, that's the case with planet earth, the ecosphere.

Stan Rowe isn't the only ecologist to take an outside-in view. The British inventor James Lovelock, who advanced the well-known Gaia hypothesis, thought from the outside in, too. (His primary conceptual error is that he regarded the earth as a super-, not supra-, organism.)

This downward view from ecosphere to ecosystem to organism is consistent with the idea of the great and late ecologist H. T. Odum. In his scenario on the origin of cells he recognized that billions of years before there were already conditions necessary for organisms to exist, material cycled and energy flowed. Sunlight would penetrate the near surface of a body of water such as the ocean, a three-dimensional region that we would one day call a photochemical zone. Materials we call nutrients are stirred by wave action. Energy in the form of sunlight allowed the formation of chemical mixtures of compounds called polymers. With continual nutrient cycling and more energy from the sun's rays, various molecules are added, making larger polymers, which eventually break apart with more additions to come.

More recent widely accepted speculation has the origin of organisms occurring near the deep thermal vents on the ocean floor. Here the energy source is from the earth's interior rather than from sunlight. This is not a detail for this discussion. The larger point is that in ecosystems everywhere, energy flow and nutrient cycling are the basic necessities for our ecosphere's myriad organisms to exist. Primitive ecosystems required accommodations to the basic necessity of nutrient cycling and energy flow, basic realities of ecosystems on the land, in fresh water, and in the sea. This primitive reality adds weight to granting priority to the ecosystem as the primary conceptual tool if we are to wisely manage what is responsible for our own livelihoods.

We pervert this reality—though we don't escape it—with our industrial applications when we use fossil energy to supply inputs such as nitrogen and other essential elements, because the integrity of a natural ecosystem has been destroyed with the tillage equipment or toxic chemicals. Rather than move our thinking to the ecosystem to explore what the ancient processes of the wild might offer, we retreat to the other end of the spectrum and rely on our faith in reductionism. For example, we tinker with the DNA by injecting genes into our crop for herbicide resistance. The soil resource with its ancient integrities has been compromised by the toxic product of the chemical industry.

The practicality of this view appears from Rowe's perspective. "Once we look inward from outside the ecosphere," we see that *biology by itself is incomplete*. Organisms do not stand on their own. They evolve and exist in the context of unified ecological systems that confer these properties called life."

CHANGING OUR WORLDVIEW

If organisms alone meet the definition of life, the physical world swirling around them then becomes part of "a life-support system." This misconception has consequences more devastating than the incorrect belief held by the Ptolemy advocates who believed that the earth is at the center. The danger of ignoring our embeddedness in this ecosphere becomes apparent as we listen to those who believe that there is another planet or moon close enough for us to live out our lives as we do here on earth. One college graduate, a successful businessman and Kansas state senator, when discussing the need for population and economic growth to end, felt it necessary to remind me that I was not "thinking galactically." We all know the universe is huge, that some scientists estimate a hundred billion stars within each of a hundred billion galaxies of the observable universe. Given the great distances between our star and others, I asked him to imagine finding a near or close enough one-to-one match for earth's gravity and its diurnal cycles and to imagine finding an atmosphere 78 percent of which is dinitrogen, 21 percent oxygen, and that will support our enzyme systems, which are keyed to other organisms. He was not moved when asked to consider that the ions in our blood set by our ancient seas will need to be accommodated on another planet. He is not alone. All of this and much, much more are taken for granted in these nondiscussions. It did not matter that the Ptolemic astronomers and the rest of the earth's human population were wrong in their time. But today, given our numbers, given our science, to believe that our destiny is in any star but our own, even one five light-years away—well, it is time to move on.

Few scientists have trouble acknowledging that it is the physical, *non*biological world that gave rise to organisms in the first place—the atmosphere, water, and the earth's crust. They can accept that organisms did not give rise to that nonbiological world. However, we *all* give shape to this world, this ecosphere. It is total interpenetration between part and whole. In addition, and perhaps more seriously, to see the physical world as "nothing but" has caused us to, as Stan has said, regard it as "loose stuff lying around that we play fast and loose with." The negative consequences of that perception are all around us, literally so. Climate change is a consequence of our disregard for the nonalive atmosphere as part of a living whole, especially during our fossil-carbon-burning age. This disregard now threatens numerous species with extinctions and unprecedented migrations. The leap from the physical world to the nonanimal and plant world, both of which we regard as nonsentient or nonfeeling, is a short one. The Amazonian rain forest is more than a swarm of diverse organisms, more than wild biodiversity we can safely sacrifice to accommodate soybeans or pastures for cattle. Our dominant insideout worldview accommodates our short-term economic imperative. Deficit spending of the earth is a derivative.

Part of this psychology is part and parcel of the history of biology. Imagine a biologist standing at a major European port during the last half of the eighteenth century or anywhere in Europe or North America in the nineteenth century. Ships are arriving from around the world. When they dock, now and then one of them has a collector or naturalist and perhaps a few assistants on board who come down the plank and are greeted by some museum staff, rich patrons, and other biologists. Or a globetraveling geographer has just disembarked. This geographer is not a mere mapmaker, for when the cargo is placed on the dock and a few of the crates pried open, the variety of homebound botanists, zoologists, and entomologists behold creatures they have never seen. A small part of the cargo carries live vertebrate creatures already heading off to be fed, watered, and sent to zoos. A few live plants are bound for gardens and greenhouses. Most are pickled, stuffed, or, if dried plants, destined to be placed on herbarium sheets. The insect collection will be stored in wooden drawers or remain in bottles. Everyone shares a fascination in these fellow organisms, but who among us would have further wondered about the various forces, both biological and physical, especially physical, that shaped those specimens? Who among us would have been thinking from the outside in? Not many of us, though clearly Charles Darwin was one.

But there is more. Those ships arriving in the eighteenth and nineteenth centuries and the professionals who pondered them were descendants of those ancestors in Europe who gave us the Age of Enlightenment, who around 1600 gave us a way to know. We often start with Nicolaus Copernicus, then move to Galileo, Francis Bacon, René Descartes, Sir Isaac Newton, and a host of others. For the past four hundred years or so we have operated with the belief that a major way of knowing is to break a problem apart, to be reductive, placing priority on the part over the whole.

Stan Rowe acknowledges this history, but he peers even further back, to the time when there was a widespread belief "in the existence of universal orders of organization," which were regarded as more important than individual organisms. He cites the Greek theory of natural science, from Plato to the Stoics, whose worldview carried over to Romans such as Cicero. Leonardo da Vinci was a late holdout of this outlook, but it was mostly plowed under during the Middle Ages. A few remnants survived to grow seeds among the nineteenth-century Romantics in Europe and North America, who in turn provided the philosophical framework for a growing number of modern conservationists, including Stan Rowe. They are mostly in the common rather than formal culture, and so aren't named in the history books. They represent a parallel to those discussed by Wendell Berry earlier. In the formal culture it is the likes of John Muir, Aldo Leopold, and David Brower.

The deck seems stacked against changing our worldview. In our schools, in our streets, in our boardrooms, and on Wall Street, in our ordinary comings and goings, the Enlightenment thinking that set the stage to accommodate the machine and the technology first smothered and later displaced the old rooted-in-the-earth symbols and systems.

But it is fair to ask what is so appealing about a reductive way of operating. What is so seductive about the mechanistic worldview? It is easy enough to acknowledge that mechanical things, especially those that use energy from fossil fuels, grant us great power at our fingertips. Travelers on the freeway with forty-mile commutes or stranded in traffic can still conduct business with their cell phones. A reductivist, Cartesian science has made this raw power possible—at least for now. This way of thinking has been adapted in physics, engineering, and medicine. "It just seems natural," we say. "Humans are lazy," we say. "It is easy," we say. "Why not follow the path of least resistance?" we ask. In our research, why not avoid hopelessly complex disciplines like ecology? Nobel laureate Peter Medawar once said, "Successful scientists tackle only problems that successfully yield to their methodology." The problem is that the methodology that gave us power and gadgets also gave us resource depletion and global warming. It is not a methodology that will ever place the whole over the parts.

Stan Rowe was asked, "Can the whole be greater than the sum of the parts?" "That question alone," he replied, "gives the game away. The question says, in effect, that 'we know the parts exist, now what about their sum?' The rightness of reduction is assumed by questioning whether anything other than parts can really exist. Science does not entertain the awkward possibility that reality might be distorted by giving priority to parts over the whole."

An increasing number of scientists, like Stan Rowe, are acknowledging that we live in a world in which books abound telling us that the individual is more important than social growth, the person more important than the world, and the fetus more important than the mother. Any organism, we are told, can be computed from the complete sequence of its DNA. The question becomes, how do we counter this perspective so dominant in nearly all societies?

Stan's inescapable conclusion is we start acknowledging that our Cartesian heritage stands in our way because it lacks coherence, which is to say, it is a *fragmented* perspective. He writes, "*the search for meaning at lower and lower levels of organization blunts the higher level search for more inclusive realities.*" This is not just another way of thinking. For in Stan's view, and now mine, this one-way vision threatens the future of the human race. It blinds us to the surpassing importance of *supra-organismic realities*—the earth's sustaining ecosystems, the planet's skin, the ecosphere. That said, we have to acknowledge that more than the teaching of Descartes is at work on us.

I thought I knew what life was, especially when I stood in the delivery room hearing the first cry and watching the first squirm of my children, or at the funeral of a family member or friend. It is easy for all of us to feel that we know life when we see it. *It's natural.* It is common sense. It will require an effort beyond the repudiation of Descartes to overcome this commonsense view. It seems possible since most members of our species once had the commonsense view that the sun came up in the east and set in the west, that the sun's rising and setting were the sun's movement, not the earth's. *It was natural* to believe this. But who believes it now? More importantly, it doesn't hurt much if someone does believe the sun is moving around the earth, but it does matter if we continue to believe that the individual is more important than social growth, the person more important than the world, and the economic growth more important than the wild ecosystems seriously under siege that "hold answers to questions we have not yet learned to ask."

Many scientists, and not just biologists, have been serious about answering the question "What is life?" When the question goes long enough, we are forced to define the terms used in the discussion. Stan Rowe insists that to conflate *organisms* and *life* is a mistake. He continues, "*But just as the living parts of an organism depend on the vitality of the whole, so living organisms depend on the energetics of planet Earth from which they evolved and by which they are maintained. From an ecological viewpoint, planet Earth, the inclusive supra-organic ecosphere, is a logical metaphor for Life.*" Looking downward from outside the ecosphere, we see that organisms do not stand on their own. They evolve and exist in the context of *unified ecological systems* that confer those properties called life.

As organisms ourselves, whether afoot in the fields, at home, or traveling widely over the globe, our human instinct is to notice other organisms, plants and animals, from near and far. As Stan says, it is natural to be fascinated with the liveliest components of ecosystems—species, populations, and communities—making it easy to be distracted from the larger realities, and preventing us, he says, from seeing the forest ecosystem for the trees. Moreover, there is the rooted and dominant conviction that plants, animals, and especially people are the most important entities, rather than the *globe's miraculous life-filled skin*. Species attract far more attention for biologists than the spaces that envelop them. And then comes Stan's damning conclusion: "It is from this error that the whole world suffers."

59

THE LAND INSTITUTE

History beyond the thinking just described stands behind The Land Institute's beginnings and current efforts. The civil rights movement and the Vietnam War were still fresh in our memories. Who among us old enough could fail to remember 1963? In September, a dynamite bomb went off in the basement restroom of the 16th Street Baptist Church in Birmingham, Alabama. Four young black Sunday-school girls were killed.

Close to the same church, a media photograph caught the image of a police dog lunging at a young black boy.

Bull Conner became a household name in that time as an agent of suppression. Eighteen days before the bombing was the march on Washington where Martin Luther King Jr. stood on the steps of the Lincoln Memorial to deliver his famous "I have a dream" speech. In November 1963 in Dallas, President Kennedy was assassinated.

The storm had been gathering that year. In April Martin Luther King's "Letter from the Birmingham Jail" was smuggled out and into the hands of an editor at *The New York Post Sunday Magazine* who first published excerpts. Many months lapsed before it was published in full.

My oldest daughter's second birthday was five days after the bombing. My son had been born in February. And then there was the spector of nuclear war hanging over us. My wife and I, along with countless others, as young parents wondered what sort of world we had brought our children into.

The summer of 1963 our family drove from Kansas to live in a dormitory in Raleigh, North Carolina, for a four-week course. It was our first taste of southern living.

A growing awareness of civil rights issues and U.S. involvement in Vietnam was electric. Kennedy had told us that we would be out by 1965, but he had also told us in his inaugural, using those Ted Sorensen phrases, that we would "go anywhere at any cost," words that Lyndon Johnson felt he had to live up to if he was to honor the legacy of his slain predecessor.

As the Vietnam War escalated, protests increased. Medgar Evers of Mississippi was assassinated; civil rights workers in the South were murdered and buried with a bulldozer; students at Kent State were shot and killed for exercising their right to protest.

Rachel Carson's *Silent Spring* had appeared in 1962. When I finally got around to reading it, there was now one more reality about what was wrong with this country. Beyond being racists, we *feared communists* and went to war, and now our country was *polluted from the chemical industry*. Like inside, like outside. Our country took the herbicide Agent Orange to Vietnam to make war against nature as we set out to destroy the vegetative cover for the Viet Cong and North Vietnamese.

The month my family and I spent in Raleigh that summer of 1963 convinced me to return to study genetics at North Carolina State in the fall of 1964. We spent the next two and a half years there. Graduate studies and research kept me from being anything more than a very marginal civil rights activist. Out of curiosity I attended a Ku Klux Klan rally with a fellow graduate student in genetics from Rhodesia, and the same day we both heard Martin Luther King Jr. His speech inside was delivered to a large assembly of mostly black people. I will never forget parts of it or the emotional power it had.

Watching the Klan take over Raleigh after the assembly, when Klan members with their long flashlights chased blacks out of public places, most other North Carolinians, it seemed to me, were as appalled as I.

But Monday came and I returned to my studies, returned to tending my greenhouse plants that would get me a dissertation. I returned to my wife and two small children for succor.

But the Klan and King and later the Black Panthers and Stokely Carmichael and H. Rap Brown and Gloria Steinem would change my world and become part of the intellectual architecture of The Land Institute.

First I returned to Kansas Wesleyan, taught there another four and a half years, and accepted the challenge to organize the Environmental Studies Department at California State University in Sacramento. My two earliest colleagues were Angus Wright, a newly minted Latin American historian, and Charles Washburn, from Mechanical Engineering. Together we faced the challenge of defining "Environmental Studies," which included writing the catalog. Wright, in his twenties, and Washburn and I, in our early thirties, were not about to let environmentalism be restricted to teaching students how to mitigate environmental impact. The civil rights movement, the antiwar movement, the gap between the rich and the poor were widening. We three young white males wanted to make important differences in our work. We couldn't be a King or a Black Panther. None of us would be a Rachel Carson, a heroine of ours, but we were not about to allow environmentalism be defined in a narrow "clean-up-the-mess" sort of way, because we, in fact, did see that war, racism, poverty, the growing gap between rich and poor, destruction of our environment, and consumerism were one subject. So we built a curriculum. It began with what Angus called our "ain't it awful" course, followed by more specialized courses. We never felt that we had included all the necessary considerations, but we made sure that the curriculum was not restricted to wilderness issues and conservation or simply pollution or the exploding population. We talked about worldviews, and we did not pretend to embrace that hopelessly naïve view that it was possible to be objective. We had a sense of oughtness, and we white boys took our stands where we could.

I became a tenured professor there, and so I could have stayed. I thought that universities were still too much a part of the problem. I wanted a world that included the physical and the intellectual. I wanted the abstractions to have particularities. I wanted to come home to Kansas, wanted to do what some then called a homesteading trip.

And so The Land Institute includes the experience with Angus Wright and Chuck Washburn and many others. It is a product of the Vietnam War, a product of the civil rights movement, and a product of my agrarian parents and rural community, and of course, of all the people at The Land over the years, including my then wife, Dana, and young family.

I mention the history in part because over the years I have detected an absence of sense of history in environmentalism. I worry that without this, our thinking will not be bold enough.

The values of a culture penetrate everywhere. The genotypes of the crops and livestock have Chicago Board of Trade genes, computer genes, fossil fuel wellhead genes, ensembles of genes that would not be there without the imperative of the Chicago Board of Trade or fossil fuel wellheads or computers or food being a commodity.

Every pollen grain and every plant egg that is fertilized carry the values of the plant breeder, and though it may seem like a stretch, every population now, a hundred years from now, and beyond will carry some "little" memory of what that violence stood for in the lifeless bodies of four young girls in the rubble of the 16th Street Baptist Church and the implications of the use of industrial-chemical Agent Orange in a faraway land on an ecosystem that was not ours to poison.

CHAPTER THREE

EARTH IS ALIVE

An alternative to the view that organisms possess "life" is that "life" possesses organisms. By this hypothesis, the secret of "life" is to be sought outwardly and ecologically rather than (or as well as) inwardly and physiologically.

—J. Stan Rowe

The practical importance of an inclusive definition of life has been well stated by Gregory Bateson:

Now I suggest that the last hundred years have demonstrated empirically that if an organism or aggregate of organisms sets to work with a focus on its own survival and thinks that is the way to select its adaptive moves, its "progress" ends up with a destroyed environment. If the organism ends up destroying its environment, it has in fact destroyed itself. . . . The unit of survival is not the breeding organism, or the family line, or the society. . . . The unit of survival is a flexible organism-in-its-environment.

We have already advanced the view that "life possesses organisms" is not in the realm of common sense. If we accept Stan Rowe's assertion that this living earth, this alive ecosphere that includes water, the atmosphere, the earth's crust, and more, is beyond people and larger than what most of us refer to as life, how do we advance the argument? We can continue with his extension: it is larger in *time* (it was here before we were), larger in *inclusiveness* (we are included within it), *more complexly organized*, and superior in *evolutionary creativity* (it gives rise to species, whereas we mostly only modify, through selective breeding, a few of the species the ecosphere has provided) and has greater *diversity* (a product of evolutionary creativity). It is implicit within Gregory Bateson's above assertion that this is not just another way of looking at the world. Stan Rowe ups the ante by asserting that nothing is more important than comprehending the overarching supraorganismic reality we call nature or the ecosphere. With this as a given, the task of ecology becomes central as we contemplate our place in the world.

We have a lot of learning and teaching to do. We have little trouble being mindful that our star, our sun, provides radiant energy to our ecosphere. Far less apparent is the reality that the heat from within the earth is a factor, too. It has been fueling the geologic activity that has pushed up and recharged the supply of organism-dependent nutrients. The soil's nutrients are pulled downhill relentlessly as runoff toward the sea and/or downward in a process we call nutrient leaching of the soil to depths beyond the reach of plant roots.

I began to think about this geologic activity during what I now call "My Field Trip over the Plates." Most scientists whose work or recreation includes field trips have little trouble telling stories of their favorite or most memorable excursions. The one that sticks out from all the rest for me lasted three days in September 1985, near Comptche in Mendocino County, California, about four hours north of San Francisco by car. Hans Jenny and Arnold Schultz led Stan Rowe and me up and down the ecological staircase of Mendocino County. The treads of this staircase are due to the Pacific tectonic plate sliding under the continental plate, pushing it upward along the coast. Waves nick away at the landmass over time, leaving extensive terraces. There are five such terraces for this story.

Beginning at the ocean's edge, the first terrace is about 100,000 years old, the second 200,000, the third 300,000, the fourth 400,000, and the fifth 500,000. Terrace one features grassland. On terrace two we see lush strands of redwood and Douglas fir. The third terrace is a transition zone with some bishop pine coming in. The fourth and fifth terraces support what is called pygmy forest.

Before I started up the staircase, based on my experience on the prairies and in deciduous forests, I was a firm believer that any natural ecosystem was sure to improve with time if the basic soil minerals essential to organisms were there. By that I mean plant life then has what is necessary to capture the other four atmospheric elements necessary: carbon, oxygen, nitrogen, and hydrogen. Such a system would become stable; a variety of creatures would find a home there. A dynamic equilibrium would be established. It was Ecology 101, which meant to me, at least, it would improve, relentlessly so over time, indefinitely. By the time we headed back toward Berkeley in the car, the pillars of my ecological understanding had been shaken. My midwestern and Great Plains perspective—and I must add what I had thought I had learned in the classroom—was seriously limited.

My concerns grew over the next several weeks. I corresponded with my fellow biologists, the three ecologists, about various matters seen and discussed on that trip. About four months after the trip, there came a letter from Hans Jenny saying that he was not aware that there was a concept of steadily improving ecosystems. I was embarrassed by my ignorance when he said that such a "sunshiny belief rests on neglect to appreciate the soil as a dynamic—either improving or degrading—vital component of land ecosystems." There was little comfort in the fact that I had had it half right.

In that same letter he wanted to know whether he and Arnold had presented Stan and me with "sufficient physical evidence that the decline in soil and vegetation from the redwood–Douglas fir forest on the second terrace to the pygmy forest is a *natural sequence*." Plant ecologists had, after all, designated the redwood-fir forest a climatic climax (from climate) and the pygmy forest an edaphic climax (from soil). In Hans's view, ecologists had designated two different worlds, "not realizing that the two ecosystems might be on the same time arrow, merely separated by a long time interval."

In comparing the luxuriant redwood-fir forest to the pygmy forest, Hans insisted that "nature might call it a biological improvement, an adaptation of vegetation to a changing substrate." Again, downward leaching of nutrients in the soil or lost toward the sea is the rule. Of course, the reality is that here and there our earth keeps experiencing various geological events, such as volcanoes, mountain formation, and glaciers, mostly in geologic time. In Mendocino County the shifting plates recharged the surface with elements necessary for organisms, which, once combined with the inorganic world, can make soil. But the relentless leaching occurs on that coast and, by the time 400,000 years have passed, we have a pygmy forest.

Europe and North America have had, in recent geologic time, uplifts. India is slipping into Central Asia. Uplifts recharge the minerals, and over the last two million years the grinding ice of the Pleistocene glaciers pulverized the rock, releasing those essential elements for organisms. Once those elements were set loose in the biota, nosing roots would capture them to combine with their atmospheric relatives—carbon, hydrogen, oxygen, and nitrogen. We in the United States are major beneficiaries of this ice, which came and went over that short geologic episode. The ice scraping away useful-to-the-biota elements from the Canadian shield made the United States the richest agricultural country of the world. We are fortunate; Australia is not. Its last geologic activity was sixty-five million years ago, and that continent is likely to have a relatively poor nutrient base for a very long while, likely beyond human time. The net primary production there, overall, will be small. Australia needs a geologic recharge event.

And so there we have it. The net primary production may decline, but the living ecosystems adjust to reduced nutrients as well as respond to their increase. These nonorganic geologic forces are part of the creative, living earth. The challenge for *Homo sapiens* is to learn to live within the means, not exceed the natural recharge rate of the forces at work on the earth's crust. Humans almost everywhere are rapidly depleting both the soil quality and quantity in agricultural time (the last ten thousand years) that the forces of the ecosphere made available in geologic time. As Hans Jenny said in his last letter to me: "The picture of natural decline of native ecosystems, more dramatically displayed by bare laterite crusts, has broad philosophical implications. Many popular writers contend that if our society were to adhere to ecological laws we would have paradise on earth, a simplistic view. The laws they cite, for example, that diversity creates stability, may not be broad laws, and maybe there aren't any laws, they are trivial. We may want to look into that."

We may want to look into that, indeed!

I shared the exchange of letters with my friend John Cobb, a

Whiteheadian philosopher. "Where do we begin?" I asked. "What can nature tell me?" In response he wisely stated, "We cannot learn from [nature], except as we ask questions and we have to be ready to have the questions revised by the answers." I wrote Wendell Berry and pulled him into the discussion. Wendell Berry agreed that we need a conversation with nature. This view favors a highly interactive approach in which, as John added, "We neither try to impose our categories nor merely adapt to what is." It is clear that if we are to make it here as a member of the biota, we need to have this conversation with our alive, creative earth.

This book is predicated on the idea that the earth is alive, which informs the spirit in which we ask the questions. That spirit and acting on it ultimately will determine our fate. Our first question at any place on earth may be, as Wendell put it, "What was here?" meaning what was the ecosystem like before disruption by humans? His second question is, "What will nature require of us here?" Rather than ask that question, we mostly ask a childish question: What can we get away with? There is no assumption here that nature has any moral authority. But it does not rule it out either. The idea that we will probably never know enough to know is a source of humility.

THE GENIUS OF THE PLACE: TWO EXAMPLES
A Nebraska Prairie and a Brazilian Rain Forest

Our earth, our ecosphere, our supraorganism, has created a mosaic of ecosystems worldwide. Each one represents "the genius of the place."

Consider two ecospheric geniuses. The first, a Nebraska prairie. The second, a tropical rain forest, let's say in Brazil.

William C. Noll, a graduate student at the University of Nebraska, conducted his research for a master's degree from September 1933 until

September 1934 near Lincoln. Noll compared an upland prairie with an adjacent field of winter wheat.

As it turned out, that eleven-month period of the study was the driest and hottest ever recorded for the area. Precipitation was below normal every month except December. For several decades, the average for the area in an eleven-month period was a little over twenty-four inches. But in the time of Noll's study only ten and a half inches fell. The soil type (Carrington silt loam) and drainage conditions for both the wheat field and the prairie were the same.

The differences in response were dramatic. Nearly 9 percent of the water that fell on the wheat field ran off. A little over 1 percent ran off the prairie. Shallow-rooted plants died on the prairie. Those of moderate depths suffered, but survived. Deeply rooted species functioned what appeared to be normally. Most of the prairie plants did survive. The wheat plants, all dwarfed, died in early June before the half-filled kernels could ripen.

It was not only dry that summer, but also hot. During the first week in June, the maximum weekly average high temperatures were 100° F for the prairie but 102° F for the wheat field. During the spring and summer, the average weekly humidity was 20 percent or lower. Humidity was higher in the wheat field but only during the short period until the wheat dried. From then on humidity was usually 5 percent higher or more on the prairie than in the wheat field.

Wind speed was affected by the landscape. There was 2.24-mile-per-hour greater wind movement on the prairie until harvesttime. Afterward, however, it was from 0.25 to 5 miles per hour greater in the field. Before harvesttime, evapotranspiration was greater on the prairie; thereafter and for a much longer period of time it was greater in the field.

The maximum surface temperature at 10:00 a.m. was 107° F on the prairie and 111° F in the wheat field. Even at three inches in depth, in July the maximum soil temperature on the prairie was 104° F but 112° F in the field. Winter and summer extremes of temperature were greater in the field to a depth of three feet.

The water losses per square foot per day as measured from phytometers after the first of April were 2 to 2.5 times greater in the field, where a maximum of 1.97 pounds per square foot was reached. Less water was used by vegetation while producing a pound of dry matter in the wheat field—2,400 pounds—compared to 5,584 pounds on the prairie. Of course, once the wheat stalk was dead, it no longer required water. The leaves and stems of the prairie plants were still alive and in need of water. But the prairie had saved more water, too. The water losses from March 13 to June 11 were 20.8 tons per acre per day in the field and 13.3 tons on the prairie.

The prairie was the area's local genius when it came to receiving, storing, and allocating water. In such a manner essentially all of nature's vegetative structure survived.

Another local genius, a Brazilian tropical rain forest, acts in a way opposite to the prairie. It has evolved to get rid of water fast. It keeps it from the soil and discharges it to the atmosphere and in so doing helps regulate the hydrological cycle of the ecosphere. It is little wonder that nearly all agriculture efforts in wet tropics lead to reduced fertility precisely because water moves nutrients downward.

Whether it is a Nebraska prairie community or the tropical rain forest or anything else in between, problems arise because *Homo sapiens* the homogenizer tends to invert what nature does well. If nature is our standard of measure for sustainable agriculture, then we should expect a mosaic of agronomic arrangements across our land. It must include, in the words of the biologist John Todd, "elegant solutions predicated on the

uniqueness of place." The creative ecosphere "acknowledges" the uniqueness and has created a mosaic of ecosystems worldwide.

The local genius in Nebraska and the local genius in Brazil as subunits
represent creative adaptations within the ecosphere. Conversely, when
humans plant soybeans in Nebraska *and* Brazil, not only is the ecological capital of both places dangerously reduced, but also long-term options
for humanity generally are reduced because both the prairie ecosystem
and the rain forest ecosystem are more than mere containers of interesting
ecological phenomena.

The interpenetration of part (in this case the ecosystem) and whole
(in this case the ecosphere) is such that whole influences part even as part
influences whole. The rain forest is a major factor in regulating the hydrological cycle of the ecosphere.

That is looking upward in the interpenetration. Looking downward
from the ecosystem is the biota, the physical components that sustain it,
and the "vert" spaces where gases exchange, insects, birds, and bats fly, and,
to show my bias, pollen is transferred.

CHAPTER FOUR

THE 3.45-BILLION-YEAR-OLD IMPERATIVE AND THE FIVE POOLS

W e humans may think of ourselves as more important than other animals. I am certainly not going to take on as a project reversing that belief. Even so, it is important to recognize the commonalities we share with other creatures since our limits and our fates are tied to so much that is common. We ingest food. We eliminate. We mate. We have young. We grow old. We die. Easy enough to internalize with a little thought, but there is a deeper commonality not widely considered that includes plant life, microbes, and most other living things. For the past 3.45 billion years nearly all life forms that we have observed have depended on energy-rich carbon as a fuel source. Either cells get this energy-rich carbon or they die. For us aerobic forms, which is to say oxygen-dependent creatures, it is one of the oldest drills. Oxygen travels from the lungs into the blood, enters a cell, comes in contact with an energy-rich carbon molecule (a sugar), and energy is released and made available for work as carbon dioxide and water leave the cell. These wastes go into the bloodstream, then to the lungs, and finally we exhale. The next breath in brings in more oxygen, and so it goes. Fruit flies do it this way. Aerobic bacteria are small enough to avoid the complex mechanics, but they do it, too. Elephants and whales, mice and men, essentially all of us do it. We do it or die. The study of physiology explains all of these goings-on.

This process is similar to the process of ingesting food, but our "hunger" for energy-rich carbon doesn't stop with food. Fossil fuels—products of photosynthesis from ages past—give us synthetic fibers, fertilizer, pesticides, and the ability to make automobiles. Even our nuclear power plants could not have been built without fossil energy. In the interest of what we might call comfort and security, we seem to lack the ability to practice restraint at using whatever energy is available. Why this is so has something to do with our lives shaped during an ancient-beyond-memory past. The outside dimension for humans was modest at first and mostly restricted to clothing and shelter needs. The cotton, wool, and wood we use also grow out of the earth and are held together by the energy in the molecules that make them up. They too are made of stored sunlight collected by green molecular traps called chlorophyll. In our Paleolithic past, we had little need to exercise restraint toward food or firewood, but now we do. Any improvement at getting these carbon-based necessities improves our chance to live. As a psychology teacher of mine once put it, "It is built into our meat." We strive to gain access to energy-rich carbon for our outsides as well as our insides.

THE FIVE POOLS

Agriculture—the First Pool

Our very being was shaped by a seamless series of changing ecosystems embedded within an ever-changing ecosphere over hundreds of millions of years. Its ability to support humans into a distant future was not on the line. The context of our livelihood kept our numbers more or less in check. Diseases killed us. Predators ate us. Sometimes we starved. The context that had shaped us was the context within which we lived. Apparently we ate grains, but without improving them, for centuries. But something happened some ten millennia ago called the agricultural revolution. It also became a treadmill. It happened first in one of these ecosystems, most likely in the land to the east of the Mediterranean, but soon spread. Hunter-gatherers initiated what would be recognized later as a break with nature, a split. This new way of being began our escape from gathering and hunting as a way of life. To set the record straight, Eden was no garden and our escape only partial. Where we planted our crops, we reduced the diversity of the biota. The landscape simplified by agriculture locked our ancestors into a life of "thistles, thorns, and sweat of brow." We became a species out of context. It has been said that if we were meant to be agriculturists, we would have had longer arms.

No matter how unpleasant this agricultural work may have been, the food calories increased. Our numbers rose; more mouths needed to be fed. No matter that they disliked thistles and thorns and sweat of brow: our ancestors loved their children and their own lives, and so they kept doing it. They had to eat. Some gave up agriculture when they had the chance. The introduction of the horse by the Spanish allowed some of the Native Americans to return to hunting and gathering, for a short while. Eventually the draft animals, especially the ox and the horse, were domesticated. These creatures used the stored sunlight of a grass, shrub, or tree leaf and transferred it to the muscle to pull a plow or bear a load. They became "beasts of burden."

This step onto the agriculture treadmill was the first toward the current and looming problem of climate change. It was in that time that humans began a way of life that would exploit the first of five relatively nonrenewable pools of energy-rich carbon—soil. Trees, coal, oil, and natural gas would follow as additional pools to rob from. It was here that we began to accelerate the breakdown and waste of what Amory Lovins once called the "young pulverized coal of the soil."

Our crops and we—both of us—were beneficiaries of the energy released as nutrients stored in the carbon compounds in the soil now became available. More will be made of this later, but for now it is enough to say that it was agriculture that featured annuals in monoculture instead of perennials in mixtures where the split with nature began. And so it was at this moment that the carbon compounds of the soil were exposed to more rapid oxidation. Carbon dioxide headed for the atmosphere, and the nutrients formerly bound up in those organic compounds—nutrients such as phosphorus and potassium—were now available for uptake by our annual crop plants. So, this wasn't really a use of the energy-rich carbon in the sense that we were after the energy stored in the carbon molecule. Rather, the breaking of the carbon compound at work in the soil was a consequence of agriculture. With agriculture, the soils that had once safely absorbed the footsteps of the Paleolithic gatherers and hunters and their food supply lay vulnerable. The hoe, along with the power to domesticate plants into crops and wild animals into livestock, turned these people into the most important revolutionaries our species has ever known. They plunged ahead in this new way of life, repeatedly modifying their agrarian technique as they went.

How many were aware they were at the forefront of a way of life dependent on deficit spending of the earth's capital? Certainly long before the advent of writing, humans must have understood that till agriculture not only simplifies the landscape but also compromises soil quality and plant fertility. Even so, the reality informed by the immediate reigned. People needed food. Energy-rich carbon molecules were the workhorses in the soil accommodating a diversity of species. The seeds from annual crop monocultures would feed the tribe. The energy-rich carbon in the grains provided these tribes with a more reliable and abundant food supply and, therefore, made possible the beginning of civilization. Eventually the descendants of these farmers had the tools necessary to expand the scale of shrub and tree harvest. Now the agriculturists could more aggressively exploit the second nonrenewable pool—forests.

The Second Pool

Five thousand or so years passed. It is easy to imagine that as the agriculturists wandered through the forests, their curious minds saw that they could cut down the forests to purify ores. This led to the creation some five thousand years ago of first the Bronze and then the Iron Age, and led to a further distancing of nature. But soon this second pool of energy-rich carbon was on its way to being used up beyond local replacement levels. This second use of carbon—deforestation—became, unambiguously, a mining operation. And it came on fast. And so the forests went down as the soils were eroding, first in the Middle East and later in Europe and Asia. And so it went for millennia, relentlessly, until recently.

The Third Pool

Only one-quarter of one millennium ago, the third pool—coal—was opened on a large scale with the launching of the industrial revolution in 1750. But already by 1700, England's forests were mostly gone to heat the pig iron. The Brits then took their ore to Ireland, where forests were still abundant, to purify the metal. The stock of the second pool of energy-rich carbon, the forests, had been so depleted that this third pool must have gladdened the heart of those who would exploit it. Coal reduced the pressure on the forests only slightly, for after the defeat of the Spanish Armada, it cost England its forests to rule the waves for the next three hundred years.

The availability of coal, this third pool, provided a quantum leap in our ability to accomplish more work in a shorter period of time. The density of energy stored in a pound of coal is greater than the density in a pound of wood. The accessibility and breakability of coal sponsored countless hopes, dreams, and aspirations of the British Empire. However, the colonialism those carbon pools made possible also destroyed local cultural and ecological arrangements that will be, at best, slow to replace in a sun-powered world.

It seems inevitable now that Neolithic farmers would move from a Stone Age and on to a Bronze Age and later, an Iron Age. Similarly, given the energy density of coal, it also seems inevitable now that a steam engine would be built to accelerate the industrial revolution.

Without soil carbon, forests, and coal, it seems doubtful that the British Empire would have had the slack in 1831 to send a young Charles Darwin on his famous voyage around the world. And once home, he was given the leisure to investigate his collections, pore over his journals, exchange letters with contemporaries, converse with his scientific peers, and finally, in 1859, have *On the Origin of Species* appear in London bookstores.

The Fourth Pool

The year 1859 was an auspicious one, beyond Darwin's publication. It was also the year of the first oil well—Colonel Edwin Drake's oil well in western

Pennsylvania—and the opening of the fourth pool of energy-rich carbon, oil. Cut a tree and you have to either chop or saw it into usable chunks. Coal you have to break up. But oil is a portable liquid fuel transferable in a pipe, a perfect product of the Iron Age.

The year 1859 was also when the ardent abolitionist John Brown was hanged at Harpers Ferry, a reality more than coincidental. In some respects, John Brown, beyond believing in the absolute equality of blacks and whites, stands alone in his time. His fervor would have received little traction had not the numbers of abolitionists been growing in the industrial North. The South had coal, of course, but not as much. It was a more agrarian society. Northern supporters, who were more profligate pool-users, could afford to be more self-righteous than the more agrarian, less coal-using, slaveholding South. Leisure often makes virtue easier.

The Fifth Pool

Natural gas has been available in some form of use back to the times of the ancient Greeks. But it did not become a manageable pool as a major power source until after coal began to be used. We count it as the fifth pool and likely the last major pool. Other minor pools may follow, such as the lower-quality tar sand and shale oil, both energy- and water-intensive for their extraction, which are in the early stages of being exploited. Over the last half century, we have used natural gas as a feedstock to make nitrogen fertilizer, which we apply to our fields to provide us a bountiful food supply while creating dead zones in our oceans. This process, called the Haber-Bosch process, was developed in the first decade of the twentieth century by two Germans, Fritz Haber and Karl Bosch. Vaclav Smil, a resource scholar at the University of Manitoba, has called it "the most important invention of the twentieth century." Without it, Smil says, 40 percent of humanity would

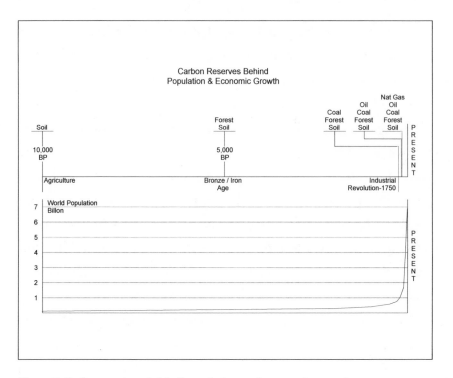

Figure 7. Carbon reserves behind population and economic growth

not be here. This is certainly a true enough statement given the reality of our cattle, pig, and chicken welfare programs.

So, in summary, a combination of the 3.45-billion-year imperative and our success in breaking into the storehouse by the discovery and exploitation of these five carbon pools is a central reality behind the "terrible truths" mentioned by Kathleen Raine. When we were gatherers and hunters, the ecosystem kept us in check. But since the advent of agriculture, we have forced the landscape to meet our expectations, and we have been centered on this way of life. We plow. We cut forests. We mine coal. We drill for oil and natural gas. We want the stored sunlight the oxygen helps release. The oxygen that enters our lungs to oxidize energy-rich carbon molecules in our cells is internal combustion—not too dissimilar to the oxygen that enters the air intake of an automobile and, with the aid of a spark, releases the energy to power a bulldozer or to run a car idling in a traffic jam.

We relentlessly rearrange the five carbon pools to get *more* energy or more useful materials. Internal combustion is the name of the game. We reorder our landscapes and industrial machinery to keep our economic enterprises (and ourselves) going, all the while depleting the stocks of nonrenewable energy-rich carbon. We are like bacteria on a petri dish with sugar.

So here we are, the first species in this 3.45-billion-year journey that will have to practice restraint after years of reckless use of the five carbon pools. None of our ancestors had to face this reality. We are living in the most important and challenging moment in the history of *Homo sapiens*, more important than any of our wars, more important than our walk out of Africa. More important than any of our conceptual revolutions. We

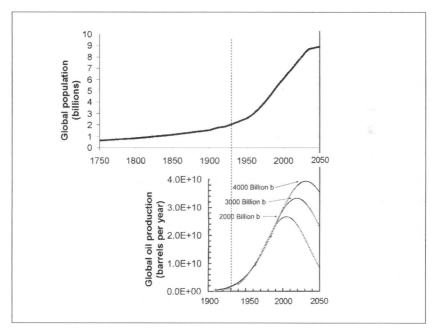

Figure 8. Global population growth and its correspondence to actual and projected growth of oil consumption in the same time period

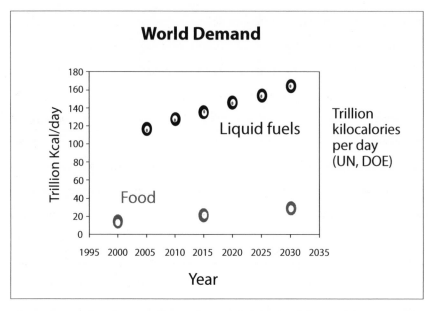

Figure 9. Actual and expected growth in global food calories and corresponding growth in actual and projected growth in liquid fuel calories

have to consciously practice restraint to end our "use it till it's gone" way of life. We have to stop deficit spending of the ecosphere and reduce our numbers if we hope to prevent widespread sociopolitical upheaval.

Figure 10. On solar-powered farms the hayloft was the "fuel tank" sponsoring meat, milk, and traction energy for draft animals. The lower story would have straw or hay for bedding, which captured manure and urine (nitrogen) to be returned to the fields via a manure spreader.

Figure 12. The ammonia (nitrogen) tank, the product of the Haber-Bosch process, made the straw as an absorber of nitrogen and the lower story no longer necessary.

Figure 11. The diesel tank, with highly dense energy, made the barn loft obsolete.

Figure 13. As a consequence of the availability of fossil fuel for traction and to sponsor nitrogen fertility in our fields, the above was not an uncommon sight in agricultural country.

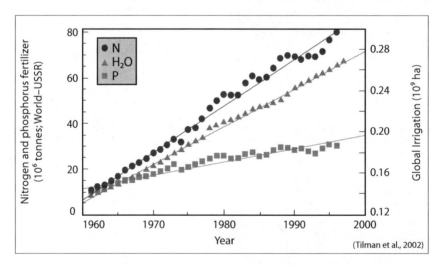

Figure 14. Life on the farm began to radically change after World War II. An expansion in the use of fossil energy made commercial nitrogen and phosphorous fertilizer cheap and available. Irrigated acreage also increased. Combined, these factors greatly increased the global food supply as the industrial mind replaced the agrarian mind.

CHAPTER FIVE

THE RISE OF TECHNOLOGICAL FUNDAMENTALISM

Millions of people go through life without knowing anything about physics. In spite of this ignorance, all the African natives do not starve; nor do all the Eskimos freeze; and the Indians, though they had none of our science, still managed to exist. But these people only exist; they do not really live. They could not live our way, for our world is entirely organized according to the principles of physics. You, yourself, must understand these principles, or you will be more or less like a barbarian in a modern city.

Without knowing physics we might still have some kind of food, shelter, and clothing; but we want a lot more than that. We want a variety of food brought from all parts of the world; we want safe houses, comfortably heated and lighted; we want different kinds of clothing for different uses and seasons. In addition, we have learned to expect quick and easy transportation by rail, airplane, boat, or motor; cheap and ready communication by radio, telephone, or

telegraph; and hundreds of other things that increase our comfort, reduce our labors, and meet our demands in the way of sports, hobbies, and amusements. All these science helps us to produce.
—opening paragraphs of *Elements of Physics*, a high school text published in 1947

S everal thousand years after agriculture began, the ancient Hebrews developed a multilayered, highly sophisticated mythology around the idea that there had been a better time but that we now live in a fallen condition. Whatever myth we assign to our break with nature, our lives in this "fallen world" have led to the exploitation of energy pools beyond the power of the ecosphere and its ecosystems to meet, or for the disturbed ecosystems to be renewed at the rate we destroy them.

Beyond energy-rich carbon pools, we look to nuclear power—ignoring that uranium is a finite fuel and that it is the use of energy that has been the destroyer. If nuclear power becomes a primary energy source, how do we keep it out of trouble? Without regulations imposing restraint, it seems certain we will use the energy to move more aggressively into the wild carbon-based biodiversity of the ecosphere, such as a tropical forest. It will give us access to more carbon-based products and further rearrange the ecosystems of the ecosphere. It is hard to keep high-quality energy out of trouble. This seems worth thinking about, for many of my environmental colleagues see nuclear power as necessary in our time. Environmental historian Angus Wright believes it will become the great divide of the environmental movement. It is one thing to ask what new information has arrived over the last three or four decades that makes it any safer today. The safety-and-storage-of-radioactive-waste problem remains. Yucca Mountain would already be stuffed full if it had been approved. And will we ever have a way of knowing when enough is enough?

We can also go on and address the "one-in-ten-thousand problem." It is generally assumed that the probability of a reactor accident, such as at Chernobyl, to be one in ten thousand years. That might seem safe enough. But with a thousand reactors, we should expect an accident every ten years on average. We currently have around 450 reactors worldwide, which means an accident every twenty-two years on average. Are we not already on schedule?

Now let's assume that we can make them totally safe during their operation and that we can solve the waste problem. Given that we now live in a world in which terrorism is on the increase, what sort of police state tactics must we employ? Such a police state would require stable governments to keep the materials out of terrorists' hands.

Finally, nuclear fusion comes up, the sort of nuclear reaction happening on the sun. Were the horrendous engineering problems ever to be overcome, fusion would be the ultimate simplifier of the ecosphere's creation.

None of these discussions mention the use of energy on a large scale as a simplifier. Agriculture was the first simplifier on the landscape, and with it species extinction and extirpation were huge. Cutting the forests added to the biotic simplification, as did coal and oil and natural gas. This has caused an information crisis: the DNA lost with species extinction is information lost to the ecosystems. The countless skills and cultural know-how for getting along in a sun-powered world that we have lost due to our irresponsibility will be slow in returning. This too is a loss of information. From the use of coal to the present, we have discarded much of the cultural and natural scaffolding for harvesting contemporary sunlight, and in its place we introduced new scaffolding dependent on fossil fuel abundance to keep it maintained. The embodied energy of the new infrastructure, whether it is cars or farm machinery, computers or cell phones, may be so keyed to the world in which they arose that many of these gadgets will become extinct as stocks decline.

TECHNOLOGICAL FUNDAMENTALISM
AND THE JEVONS PARADOX

Another unfortunate reality of this age of carbon exploitation is the rise of technological fundamentalism. We have interpreted the Platonic idiom "Necessity is the mother of invention" to mean "technology will bail us out." We hold a widespread belief that we will get more efficient. But this usually means that we will get more efficient technologically and not that we will do with less. (We will come to rely on hybrid cars or no-till farming, not do without.) We also believe that though nature's ecosystems may be destroyed, new energy sources will be found to compensate for our past and future destruction. I know one intellectual who seriously suggested we could invent soil when it is gone.

It should be immediately apparent that the very opposite is true. As the economist Thorstein Veblen said: "Invention is the mother of necessity." Wallace Stegner once wrote about the things once possessed that cannot be done without. Think of how many geegaws in our daily lives we didn't know we needed until we saw them—and then we became dependent on them. The fundamentalist tenet, the most widespread at the moment, is that *technological efficiency* will save us. Rarely mentioned is the Jevons Paradox, a reference to William Stanley Jevons, a British economist and author of a book published in 1865 entitled *The Coal Question*. Jevons argued that in industrial England greater technological efficiency led to *more* resource consumption, particularly coal and iron. Greater efficiency frees up capital for economic and, therefore, energy expansion. Wal-Mart's truck fleet gets better mileage, but what will keep the owners from building more box stores? Money, like energy, has a hard time staying out of trouble. For example, imagine someone with $25 million derived from a sale of his or her oil wells. Now imagine a Kansas rancher who owns twenty-five thousand acres in the Flint Hills featuring tallgrass prairie. Because the oilman loves to look at beautiful, never plowed, uninterrupted tallgrass prairie, he pays the rancher $1,000 per acre. Compared with most U.S. current agricultural enterprises, the tallgrass prairie ranch uses little fossil carbon, because the cattle harvest contemporary sunlight. For activities such as shipping, processing, and distribution, the fossil energy calories used per pound of red meat gain on the grass-fed beef are lower than a pound gained in the feedlot. Said otherwise, the grass-fed beef uses less fossil energy than the cattle raised in a feed lot. So the new owner's $25 million, derived from oil and spent to buy that Flint Hills ranch, would be a "carbon offset," not a complete offset, but a large one.

But the seller of the ranch will have a hard time keeping his $25 million out of rapid-climate-change trouble. Those dollars will be a magnet to attract other carbon-sponsored activity such as travel, or "stuff" such as a yacht, car, house, welder, lathe, whatever. Suppose he invests in wind machines with electric generators. That would be better, but we need to know how much fossil energy it takes to organize, build, and run that technology. More importantly, how do the investors in wind spend the profits? Not all of the $25 million will turn "green"; in a final accounting it might be very little. One person's attempts to lower carbon use can enable another's increased use.

Without a cap on carbon at the wellhead, the mine, or the point of entry, there is every reason to think that Jevons will be proved right again. Our culture needs to make plans to avoid the paradox. We need to understand, for instance, that in manufacturing, the reasonable consideration is not just the carbon *in* the factory, but also the carbon used to make the factory, to move components in and products out to market, and more.

Since we are already overdrawing the capital of the earth's ecosystems (forests, soils, water), our minds naturally leap to renewable energy sources to be exploited. But given our level of consumption, there are serious problems with renewables. The price for photovoltaics is coming down. Wind-generated electricity gets a favorable rating because electricity is energy of the highest quality. Storage and transmission will be a challenge, but both are being worked on. And so far we have been primarily an oil/natural gas society. As we look to renewable portable liquid fuels, we are now turning to ethanol or biodiesel. Will any of these make us more efficient?

Now to address the belief that we'll forever find new energy sources to compensate for future and past destruction. Sure, there are tar sands and shale and nuclear power, but they aren't nearly as efficient as handling coal, oil, and natural gas. It is worth being reminded again and again that it is not climate change that has destroyed so much of the ecosphere, it is energy. Our use of nonrenewable energy has destroyed rain forests. Our intensive application of nitrogen fertilizer on our farmlands has given us dead zones in our oceans. When we look at the few examples where an energy source has made it possible to restore what we have lost, it resides primarily in the realm of ecological restoration.

What this review reveals is the absolute necessity of conservation, where *sufficiency* has standing over *efficiency*, though, of course, we need both. Conservation of scarce materials is our largest *source* of energy, a source embedded within sufficiency, efficiency, and reduced population.

When my family and I moved back to Kansas from California, it was for the stated, comically expressed purpose of "figuring out a way to stay amused while we live till we die—cheap," meaning inexpensive to the ecosphere. The five of us did a sort of homesteading trip with garden, a milk cow, steer, and pigs. We burned wood for heat. It was somewhat like the farm where I had been born and raised. Yes, we did have a car, a pickup truck, electricity, even a fuel oil furnace, but during that period our ecological footprint must have been the smallest it has ever been. My wife, Dana, was an excellent gardener and beekeeper. The three children were terrific, mostly noncomplaining workers, with their various chores. What did we learn? It was a lot of work, and though we were aware ahead of time, we had the direct experience, beyond the abstraction, that it is essentially impossible to create an island in the context of our highly extractive culture. Success will require society to commit itself to a resilient future. I am confident there will be an increasing number of creative options ahead for all of us. And we can do it. We can pull away from the damaging shibboleths of the technological fundamentalists. We can begin to move our species into a more sustainable relationship with our natural capital. But first, as a technological species, we must come to terms with the Jevons Paradox and not count on efficiency alone, or on much from biofuels from the landscape. If we were, as a nation, to put a cap on carbon at the wellhead, the port of entry, and the mine, as well as timber harvest at a renewable level, then we would have the chance to see what technologies emerge from the market forces. Waiting until that energy is allowed to run freely in the culture creates horrendous accounting problems and allows countless bookkeepers of business to cheat even without their knowing it.

Most industrial countries—with more than adequate standards of living—use about half the energy per capita that we Americans do. This includes the sum of energy to households, transportation, manufacturing, and infrastructure. By saying yes to conservation, we are saying yes to efficiency, and have taken the first steps to saying yes to sufficiency. By saying no to our belief in technology as a primary solution, we can involve all sectors of our economy that link us to the terrible truths.

What seems to run ahead of everything else at the moment is that the consequences of increasing technological complexity, spurred by scientific discovery, go unmanaged because society can move only on the available paths, which are limited in number and design. Stated otherwise, the problems generated by the science/technology/economy trinity require solutions that would have to evolve along social/political paths that are too small or not available.

COUNTERINTUITIVE BEHAVIOR

Professor Jay Forrester was a systems theorist and analyst at the Massachusetts Institute of Technology. Four decades ago he trained a group of young scientists who published the famous *Limits to Growth* study in the early 1970s. Donella Meadows, lead author of that important volume, once quoted her mentor:

People know intuitively where leverage points are. Time after time I've done an analysis of a company, and I've figured out a leverage point. Then I've gone to the company and discovered that everyone is pushing it in the wrong direction!

Meadows explained that "the classic example of that backward intuition" was the first model Forrester's group ran. It turns out that the major global problems—environmental destruction, poverty and hunger, unemployment, resource depletion, urban deterioration—are all related. The common culprit was growth, both population and economic growth. "Growth has costs," she wrote, "among which are poverty and hunger, environmental destruction—the whole list of problems we are trying to solve with growth!"

Morris Berman, author of *Dark Ages America: The Final Phase of Empire* (2006), contends that the question before us has to do with how much flexibility we have in making choices and acting on them. Every civilization is a "package deal," he says, and the configuration of that package dictates a trajectory imposed by the constraints of that "deal." Some of those constraints may be positive, others negative, but the important idea is that they crystallize into a specific pattern or direction early. So when economic and population growth has been used to solve problems in the past, has been the "package deal" so to speak, there is little wonder that we might fail to realize that the factors that allowed our civilization to rise to power are the same factors that may do it in. Here is a problem: for our country to get where it is today required that we reject countless alternatives along the way. Berman argues that collapse or decline can only be avoided *if the alternatives that have been repressed* are incorporated into the dominant way of being. For example, the loss of local dairies, local butcher shops, and local produce through various subsidies, in favor of cheap fossil fuels, now makes us vulnerable.

Forrester once published a paper entitled "The Counterintuitive Behavior of Social Systems." He gave examples in the housing industry and highway building about how deficient housing became more so and how the highways actually led to more congestion. Could it be because our social, economic, and ecological systems are so complex that we have a limited ability to see beyond the complexity to an eventual outcome? For scientists, part of our limitation has to do with our reductive approach to problem solving. Ecologist Stan Rowe said, "Reductionism blunts the higher level search for more inclusive realities." So our very training may stand behind our failure to imagine the consequences of complexity.

I often invite visitors to The Land Institute to take a ride on a merry-go-round I rescued from an abandoned elementary school yard and installed in our yard at home. It stands ready to be ridden at any moment, but it is there for more than fun. I want my grandchildren—and visitors—to experience an illusion. The design features a centerpiece with an offset pivot to which four rods are attached, rods that reach back to handles in front of each of the four seats. A rider, by moving the handle back and forth, can cause all four seats and any rider to move around and around until dizziness (or apathy) sets in. Anyone riding that plaything and looking to the center where the rods are attached will wrongly perceive that the center and the attached offset pivot part are moving. The perception of movement will be retained in the rider's mind when stepping off. Anyone on the ground staring at the center and pivot will see that neither moves. Anyone sitting on one of those four seats is like a person riding on planet earth who, when looking at the center, perceives that the center is moving.

Four centuries ago Galileo was taken by Copernicus's idea that the sun was at the center of our world. The dominant idea that the earth was at the center and that the sun revolved around it was a derivative of Ptolemy, a geographer and astronomer from Alexandria who lived in the second century AD. After Ptolemy, mathematicians and other thoughtful people elaborated on what seemed obvious to all. They formalized a way of looking at an earth-centered universe. Some mathematicians who built on Ptolemy's ideas predicted the position of the known planets on any particular day of the week in any year. In spite of such powerful prediction, and no matter that Dante had used Ptolemy's cosmology as a map for the paradise section of the *Divine Comedy*, both Copernicus and Galileo believed that the Ptolemaic worldview and the Church were wrong.

Galileo, like Copernicus, was able to mentally step outside the system and expand the boundary of consideration. The power structure of the Church refused to take this courageous mental step. Martin Luther didn't help matters. The Reformation was under way, and the threat to Church leadership was heightened. Some defenders of the Ptolemaic view argued that God would not trick us. Eventually, of course, others became convinced of the truth of this insight, and a major tenet of science was born—*perception is not always reality.* This was one of the great moments in the early history of science. Unfortunately, we too often ignore or forget this lesson. It is easy to believe that we are more agriculturally productive than we really are. If we were to step off the merry-go-round and expand our boundaries of consideration, we would immediately see that without fossil fuel and material subsidies from the extractive economy, yields would be seriously lower. We could see that a focus on bushels and acres distorts our view of reality. Were we to eliminate nonrenewable resources from production efforts—stop using natural gas as the feedstock for nitrogen fertilizer, herbicides, insecticides, and more, for example—and force ourselves to rely on the natural fertility of the soil, introducing crop rotations featuring legumes for nitrogen fixation, fertility yields would plummet. High yields dazzle us and give us an illusion that we are sophisticated on the subject of agronomics.

It is only in the last century—the last 1 percent of our history with agriculture—that we have experienced such a major bump in food production. No comparable bump is likely to be attainable again, because of the degradation of the natural fertility of our soils, as well as the loss of some of the genetic potential of our major crops. According to devotees of the modern paradigm, it has been a great ride. Social systems, like ecological systems, far more complex than a ride on a merry-go-round, obscure the realities before us.

W. H. Auden's poem "The Age of Anxiety" includes this sentence: "We would rather be ruined than changed."

AGRICULTURE AS ASTEROID SHOWER

A laminated wall hanging in my office, thirty-six inches high and thirty inches wide, is "A Correlated History of Earth." More than four billion years are represented on this convenient chart. Column heads include tectonics, era, period, plants, invertebrates, fish, reptiles, birds, and mammals. One is labeled "astroblemes," for the asteroids that have smacked the earth and created "blemishes." There have been a few big ones and a lot of little ones. The one that divides the Permian-Triassic was the largest, more than 300 kilometers in diameter. From my reading of the wall hanging, the one that appears to have done the greatest damage was over 200 kilometers across, or 120 miles. That's the one that ended the Cretaceous period sixty-five million years ago. I read once that at the point of impact, the tail of that asteroid would have been as high as Mount Everest. How someone found that out is beyond me. For some time it appeared to many scientists that the asteroid strike alone finished off the dinosaurs, allowing our living supraorganism, the earth, the ecosphere with its embedded ecosystems, to eventually create a diversity of mammal types. As is usually the case in science, tidy stories have a way of becoming more complicated. It now seems volcanic eruptions under way between sixty-three and sixty-seven million years ago created gigantic lava beds called Deccan Traps. They covered an area more than twice the size of Texas. It seems that some 80 percent of the lava was spewed out in one episode. The point is that serious blows to the wafer-thin miraculous skin of our planet did not destroy the power of the ecosphere to create new organisms. The ecosystem processes that are disrupted by the massive extirpations and extinctions will return.

We can see now that agriculture, with its beginning ten thousand to twelve thousand years ago, has been like a shower of small asteroids striking our planet. As the earth was increasingly "peppered" by annual monocultures disrupting natural ecosystem processes, we moved on to adjacent landscapes. We are now approaching the end of an agriculture "agrobleme shower." In less than a century, the world's population has tripled, and we could add as many people in the next fifty years as the total population had fifty years ago. Over the past forty years, as we have doubled food production, we have wiped out numerous species, as well as land race crop varieties and local cultures.

If asteroids were known to be approaching earth, ready to deliver an impact, there would be considerable expenditure and heroic maneuvering by the United Nations. All would agree that the earlier we get started to intercept this "agrobleme shower," the better. A nudging at great distance would require a smaller angle of deflection. Economists might even calculate the cost-benefit ratio. The ecosphere as a supraorganism can absorb major blows and continue to create new organisms. The earth's creative capacity will continue. New opportunities for a change in the flora and fauna are always opening up. The asteroids and volcanoes may contribute to the history of this creative endeavor. It should be noted, however, that most healing would be observable only in geologic time, not a human's time frame.

PART II

Losses

CHAPTER SIX

THE MOST SERIOUS
LOSS OF ALL

The history of the extratropical grasslands . . . in the world is much like our own. Southern Russia, the Pampas, Australia, and South Africa repeat largely the history of the American West. The industrial revolution was made possible by the plowing-up of the great non-tropical grasslands of the world. So also was the intensification of agriculture in western Europe, benefitting from the importation of cheap overseas foodstuffs, grains, their by-products of milling, oil-seed meals. Food and feed were cheap in and about the centers of industry, partly because the fertility of the new lands of the world was exported to them without reckoning the maintenance of resource.

> —Carl O. Sauer, "The Agency of Man on the Earth"

In 1942 Dr. Walter C. Lowdermilk issued in mimeograph form a paper titled "Conquest of the Land through Seven Thousand Years." It was later published as a booklet and widely distributed by the newly established U.S. Soil Conservation Service, where Lowdermilk had been assistant chief. There were broad thinkers in the Department of Agriculture during the Roosevelt administration. Someone in charge— I always imagined it to be Hugh Hammond Bennett, the founding chief of what began as the Soil Erosion Service—sent Lowdermilk and his colleagues on an eighteen-month tour of Western Europe, North Africa, and the Middle East. Their chore for him was to survey lands that had been cultivated for a thousand years or more. "Conquest of the Land" was a summary of their findings.

Lowdermilk takes his reader to the early beginnings of agriculture. He describes the alluvial plains of Mesopotamia, a land once filled with great cities and dense populations, but in Lowdermilk's time, a place of desolation. Soil erosion had caused silt to fill the canals, despite arduous attempts by laborers to keep them clean. Here and there over time ancient irrigation systems broke down, and as Lowdermilk explained: "Stoppage of the canals by silt depopulated villages and cities more effectively than the slaughter of people by an invading army."

As Dr. Lowdermilk and crew traveled through regions that had supported ancient civilization, he was always mindful of what had been subtracted. From what scholars count as the earliest major recorded work of literature, *The Epic of Gilgamesh*, we can read the first available mention of two ecosystems, a grassy plain and a cedar forest. The epic was preserved on clay tablets in cuneiform and written a full thirteen hundred years before Homer's *The Iliad* and *The Odyssey*.

There are several wonderfully interesting tales in this epic. They feature the usual courage, determination, patience, fortitude, and so forth we expect of a hero, this time as Gilgamesh presses on in his quest. For our purposes, there are references to the rich biota of those two ecosystems.

Early in the epic is a description of a wild man, a hairy creature who came into the world fully grown, running with the wild animals and suckling them. According to a frustrated young hunter, he "feeds with the gazelles on the grass of the plain, and . . . drinks at the water place with the wild beasts." The young hunter complains to his father that this wild man "is robbing me of my livelihood. He tears up the traps I set. He releases the beasts and small creatures of the plain."

This wild man, called Enkidu, is set up to be seduced by a priestess from the temple in the city of Uruk. Seven nights and six days they spend together, and after this, when he returns to the wild animals, they will have nothing to do with him. He returns to the woman, apparently smitten, for he sits at her feet and looks attentively into her face. Though he has lost his ability to connect with the other animals and can no longer run as fast as the gazelle, he has gained understanding and wisdom. He is now a human and becomes a friend of Gilgamesh.

After some time Gilgamesh and Enkidu set out for the Green Mountain to find and kill Humbaba, the watchman of the forest. As they approach the entrance to the Cedar Forest of Lebanon, they are struck "*by the tremendous height of the shade-bearing cedar tress on the face of the entire mountain*" (my emphasis). The translator reports that forest to be 30,000 square miles, three-eighths the area of Kansas. It was not a woodlot. Eventually, they meet the fierce forest watchman, a giant, and they kill him. Gilgamesh's men are now free to cut trees from the forest.

102

One day Enkidu becomes ill, and as his condition worsens, he asks to see his friend Gilgamesh, who comes to his bedside in grief. We have a date on this, because Gilgamesh was apparently a real king who ruled sometime between 2700 and 2500 BC in ancient Mesopotamia. In the epic we read of his lament for his dying best friend, Enkidu:

> May the bear, the hyena, the panther, the tiger, the leopard, the lion, the oxen, the deer, the ibex, and all the wild creatures of the plain weep for you. May your tracks in the Cedar Forest weep for you unceasingly, both night and day. May the Ula River, along whose banks we used to walk, weep for you. May the pure Euphrates, where we used to draw water for our waterskins, weep for you.

Big wild animals? A large forest? Pure water in the Euphrates! This is the region of modern Iraq. Lowdermilk's narrative is mostly about soil loss and the rise and fall of civilization, but here we see the little mentioned loss of biodiversity.

Now back to the Lowdermilk team and their journey. They moved on to the Sinai Peninsula, toured mountain slopes that had become fields, which in turn had long ago experienced a rapid soil erosion rate that Lowdermilk did not expect to see: "The original brown soil mantle was eroded into enormous gullies as shown by great yellowish gashes cut into the brown soil covering." Such erosion had turned plateaus to plains, or peneplains, dating back to the Miocene.

Farther north, in Petra, the deserted ruins of a once-flourishing civilization greeted the team. Scholars had used Petra as evidence that the climate had become drier in the past two thousand years, making it unviable for large populations. Lowdermilk found contrary evidence. He blamed soil erosion so damaging it had created a land unable to support crops.

On to Syria, there they came upon erosion even more disastrous.

Three to six inches of soil washed from the hillsides and resulted in more than a hundred dead cities. To this Lowdermilk gives his most ominous declaration yet: "Now that the soils are gone, all is gone." The species inventory in the Gilgamesh epic would not ever again be so large.

Lebanon, a land previously covered with forests and fed with heavy rains, had been cleared of trees. The slopes were then cultivated, and the soils eroded. A permanent agriculture was almost impossible to establish. Walls built across the slopes proved to be useless against the rains. Some fifteen to seventeen centuries after Gilgamesh, King Solomon struck a deal with the king of Tyre to purchase cedars in order to build the temple at Jerusalem. Representatives of those are mostly now found only in cemeteries of that region.

Following the "westward course of civilization," the Lowdermilk group then traveled through Cyprus and North Africa. In northern Algeria, the ancient Roman city of Cuicul (now called Djemila), once rich and populous, was covered completely by erosion debris from the surrounding hills. Only three feet of a single çolumn could be seen. Lowdermilk laments, "The surrounding slopes, once covered with olive groves, are now cut up with active gullies."

Lowdermilk returned home to announce that the United States was on the same path of devastation. Basing his conclusion on erosion data derived from experiment stations located throughout the country, he warned Americans that clearing and cultivating sloping land was exposing soils to increased erosion by both water and wind. We are on our way to a "self-destructive agriculture," he cautioned. In our time nearly 300 million acres of farmland is still eroding at a rate faster than new soil is being formed. Migrating to new land may have been an option centuries ago. No new continents await us now.

The Lowdermilk team did not visit ancient Greece, but we do know that the still-appealing Greek landscape had sponsored a people that defined much of Western civilization for us today. They needed forests and soils to make those notable achievements possible. Now it is a land where the impact of agriculture is visible everywhere. Episodes of deforestation and soil erosion have gone on there for eight thousand years. Recent archaeological evidence of soil erosion due to agriculture is well documented by a team headed by Professor Curtis N. Runnels.

According to Runnels, the story begins with the farmers who first settled Greece when the landscape was pristine. Archaeological investigations of ancient ecosystems, using soils and fossil pollen along with human relics and artifacts, reveal that when hill slopes lose their soil, people leave; when usable soils re-form *thousands of years later,* people return to farm. Here is the land where both Plato and Aristotle witnessed land degradation and its consequences firsthand. Plato, in one of his dialogues, has Critias proclaim:

> What now remains of the formerly rich land is like the skeleton of a sick man, with all the fat and soft earth having wasted away and only the bare framework remaining. Formerly, many of the mountains were arable. The plains that were full of rich soil are now marshes. Hills that were once covered with forests and produced abundant pasture now produce only food for bees. Once the land was enriched by yearly rains, which were not lost, as they are now, by flowing from the bare land into the sea. The soil was deep, it absorbed and kept the water in the loamy soil, and the water that soaked into the hills fed springs and running streams everywhere. Now the abandoned shrines at spots where formerly there were springs attest that our description of the land is true.

Not all erosion is human-made, of course. There was erosion during the last ice age due to climate changes. The past five thousand years, however, is another story. Four episodes of erosion—at about 2500 BC, 350–50 BC, AD 950–1450, and in recent times—according to Runnels "were followed by a period of stability when substantial soil profiles formed." So it was more than the civilizations of Mesopotamia—Sumer, Akkad, Babylonia, and Assyria—that rose, fell, and disappeared. The story of soil and water abuse differs only in detail. So just as water brought salt, in Mesopotamia where wheat once grew, then salt-tolerant barley, and eventually, places where next to nothing was grown, rising groundwater now salts the fields in southern Australia. The populations and culture that moved upward in the Tigris-Euphrates Valley were mere foreshadowings of other places across our ecosphere in our time.

While the Greeks' experience of soil loss was well under way, in the benign climate of the hilly Apennine Peninsula Rome began relying on local natural fertility. Like the Greeks, Romans too worshipped nature deities and called the earth *mater terra*. The chapter on the Romans reads much like the chapter on the Greeks: when field fertility declined, their faith that human cleverness or intervention would pull them through increased. Virgil, Ovid, and Seneca were major promoters of such a view. Cicero is on record as having said: "By means of our hands we endeavor to create as it were a second world within the world of nature."

Phoenicians, Greeks, Carthaginians, and Romans, each in turn as their own land-based prosperity gave out, established distant colonies and empires. Egypt fared better primarily because the renewal of lost nutrients came from what might be considered an acceptable mining operation. As Herodotus, the Greek historian, said, "Egypt is the gift of the Nile," as indeed it is. The Nile received silt from volcanic highlands of Ethiopia, thanks to the predictable monsoon rains arriving from the Indian Ocean each year, bringing minerals into the annual floods of the Nile's tributary, the Blue Nile. Egypt prospered from an acceptable natural subsidy of Ethiopia's minerals. The White Nile contributed its organic matter, with its jungle origin and swampy places, and the best sources from two countries converged at the confluence. Downstream these fresh nutrients and organic matter combined to spill over a layer one millimeter thick each year to be turned into crops for Egyptian farmers and pharaohs. Without the steamy jungles and volcanic ash accompanied by uplift generally, no pyramids would have been built.

Daniel J. Hillel (*Out of the Earth: Civilization and the Life of the Soil*) and David R. Montgomery (*Dirt: The Erosion of Civilizations*) have both written eloquently about soil and civilization. With regard to the Nile they outline subtle and not-so-subtle consequences of the dams, particularly the Aswan High Dam, which was begun in 1960 by the Egyptians but largely financed by the Soviet Union. It took ten years to finish the construction, and now forty years beyond, with electricity and flood control, every drop of water is used before reaching the Mediterranean. Coastal erosion, subsidence, and soil salting continue on this longest river of the world, which still holds the record of the longest continuous civilization, now numbering fifty million people in the Delta alone, an area the size of Delaware. The periodic flooding has stopped, and soil compaction has become an emergent problem.

WE GET A NEW START

The old world, with its countless failures, was on the minds of our revolutionary fathers in colonial America as they drafted what amounted to a new set of plans during and after the Declaration of Independence. Enough of them were well-educated Children of the Enlightenment. Now they had a more or less clean slate on which to build a new republic. The slavery issue was more than a smudge on that slate, for sure, but the institutional structure was sufficiently in place that a different pattern of land seemed possible. Even so, when it came to doing the future business of agriculture, a major leap had to wait until May 1862. This was more than a year after Abraham Lincoln became president, and after the attack by Confederates on Fort Sumter and the beginning of our Civil War. With congressmen from the South not present, Congress was able to pass the Homestead Act. Now any person twenty-one or over, the head of a family, a citizen, or an alien who wanted to become a citizen, could "prove up" after living five years on the land and improving it. The title to 160 acres of public land was then his, and in some cases hers. Those opting to pay $1.25 an acre could avoid the residence requirement. Because public land was considered worthless unless improved, the sponsors of the idea argued that any who converted unoccupied land into farms should not have to pay. For the elected officials struggling with the budgets the public lands suggested a way of gaining money for our new government, which was broke or nearly so after the Revolutionary War.

As the natives were being pushed westward and increasingly marginalized, by the 1840s and 1850s, promoters of the homestead movement became more vocal. Several bills were introduced in Congress, and understandably, most Southerners opposed the bills. It clearly would mean more political power for the expanding population in the North. So it is no wonder that when the Southern states seceded in 1861, the Homestead Act was passed a year later.

Hundreds of thousands of settlers moved west during the thirty-eight-year period from 1862 until 1900. Imagine anywhere from 400,000

Figure 15. Scandinavian immigrants about to finish plowing short-grass prairie in Mellette County, South Dakota, about 1917 (photo courtesy of Luree Wacek, White River, South Dakota)

to 600,000 families gaining new farms and homes. Naturally, a speculator class arose, advertising opportunities for settlement, many of them containing outright lies to audiences in America and Europe.

The act disappointed most of those who supported it. Some of the best land was granted to railroad builders, who liked long stretches of flat places. The fertile valleys were prime. It wasn't just the railroads. People even resented the land granted to the states for agricultural colleges.

Who were these people, these homesteaders? Many had never farmed, and many of those who did "came with vision but not with sight," as Wendell Berry once said. On the Great Plains, for example, dry weather conditions did not suit those who had come from high-rainfall areas. So, in 1873, Congress saw fit to modify the original act. Larger tracts were offered. But it was still too dry to make a farming living for most. They quickly sold out to speculators, who bought thousands of acres from discouraged homesteaders, who returned to the wetter East.

On the return, some camped in my grandparents' pasture near Topeka, embarrassed to admit defeat, offering such explanations as "going to visit the wife's family for a spell." Locals knew they had been dried out. As a consequence, the land and mining resources of the West became controlled by fewer and fewer people. It has been estimated that during that thirty-eight-year period mentioned earlier, from 1862 to 1900, not more than one acre in six, maybe only one in nine, was actually settled by a homesteader.

So the skeleton of the Homestead Act represents fulfillment of an old demand, partly because of a bargain sale six decades earlier in which we acquired 829,000 square miles for $15 million. It was an unconstitutional act Jefferson pulled off to make way for farm families, believing they were "the most valuable citizens." It was an earlier, though clearly a different, version of Homeland Security. National security for Jefferson featured farmers as "the most vigorous, the most independent, the most virtuous, and they are tied to their country, and wedded to its liberty and interests by the most lasting bonds." Being "cultivators of the earth" made them so virtuous that democracy would flourish.

In *The Unsettling of America* Wendell Berry made clear that it was not a mere oversight when Jefferson failed to mention the nonfarmers, "the class of artificers" as he called them. He saw this class to be "the panderers of vice, and the instruments by which the liberties of a country are generally overturned." Some would say that he was an elitist, for he thought that "the highest degrees of education" should be "given to the higher degrees of genius, and to all degrees of it, so much as may enable them to read and understand what is going on in the world, and to keep their part of it going on right: for nothing can keep it right but their own vigilant and distrustful superintendence."

It is important to keep our mind on Jefferson's bias. Less than three months after the Homestead Act, this time a Yankee—Jefferson was a Southerner—Justin Smith Morrill, a representative and later senator from Vermont, introduced and got passed what became the Morrill Act. This act called for "an amount of public land, to be apportioned to each State a quantity equal to thirty thousand acres for each Senator and Representative in Congress. . . . To the endowment, support, and maintenance of at least one college where the leading object shall be . . . to teach such branches of learning as are related to agriculture and the mechanic arts . . . in order to promote the liberal and practical education of the industrial classes in several pursuits and professions in life."

So, in the thirty-sixth year after Jefferson's death, within a three-month period in 1862, we got an institutional fulfillment of an old longing in the Homestead Act, and with the first land-grant college act, the democratization of knowledge. Twenty-five years later, in 1887, the Hatch Act was added by Congress to establish the state agricultural experiment stations. Their primary purpose was to promote "a sound and prosperous agriculture and rural life as indispensable to the maintenance of maximum employment and national prosperity and security." The Hatch language continues: "It is also the intent of Congress to assure agriculture a position in research equal to that of industry, which will aid in maintaining an equitable balance between agriculture and other segments of the economy. . . . It shall be the object and duty of the State agricultural experiment stations . . . to conduct . . . researches, investigations, and experiments bearing directly on and contributing to the establishment and maintenance of a permanent and effective agricultural industry . . . including . . . such investigations as have for their purpose the development and improvement of the rural home and rural life."

So now, from 1862, for a while we had free land. We also had afford-able land-grant colleges and, a quarter century later, experiment stations at all of them. The years after 1887 continued apace. Another twenty-seven years later, in 1914, the Smith-Lever Act created cooperative extension: "In order to aid in diffusing among the people . . . useful and practical infor-mation on subjects relating to agriculture and home economics, and to encourage the application of the same."

Let's step back and scan the meaning of that half century, that fifty-two-year period, from 1862 to 1914. We had the acts we now call the land-grant college complex, where the dream of Justin Smith Morrill had been fulfilled. We know this because around 1874, Morrill had written of his intentions in a memoir. Remember, this is a full eleven years before the experiment station act of 1887 and forty years before co-op extension. In *The Unsettling of America*, Wendell adds that "Morrill was aware, as Jeffer-son had been, of an agricultural disorder manifested both by soil depletion and by the unsettlement of population." Morrill's memoir includes the fol-lowing: "The very cheapness of our public lands, and the facility of pur-chase and transfer, tended to a system of bad-farming or strip and waste of the soil, by encouraging short occupancy and a speedy search for new homes, entailing upon the first and older settlements a rapid deterioration of the soil, which would not be likely to be arrested except by more thor-ough and scientific knowledge of agriculture and by a higher education of those who were devoted to its pursuit."

Morrill explains his motives to support the artificers: "Being myself the son of a hard-handed blacksmith . . . who felt his own deprivation of schools . . . I could not overlook mechanics in any measure intended to aid the industrial classes in the procurement of an education that might exalt their usefulness."

Wendell carefully analyzes the language that distinguished the differences between Jefferson and Morrill. It is clear that the codification of Morrill represents a statement of both his values and the values of the citizens at the time. It was a watershed moment in the history of American thought. Here we see, perhaps for the first time in a policy sense, the influence science and the industrial revolution had on a higher education. Teaching only those destined to pursue the so-called learned professions was no more. Farmers and mechanics and all those who must win their bread by labor "need no longer be left to the haphazard of being self-taught or not scientifically taught at all."

Beyond the "democratization of knowledge," more needs to be said. With everything in place—land, education, experimentation, extension—what has gone wrong? Wendell notes that the thinking of Jefferson and that of Morrill were dramatically different. The more important difference was the apparent absence from Morrill's mind of Jefferson's more complex sense of the dependence of democratic citizenship upon a classic and broader education. It is clear from his plan of education at the University of Virginia, which lacked any form of specialized or vocational training. He must have assumed that communities were best established, stabilized, and preserved by the virtues of citizens who elected good leaders. In his analysis, Wendell makes clear that this left the "practical arts" to be improved as a matter of course by local example, reading, structured conversations, and intelligent discourse. Morrill looked at education from a strictly practical or utilitarian viewpoint. He believed that the primary aims of education were to correct the work of farmers and mechanics and "exalt their usefulness." In distinguishing among the levels of education, he did not distinguish, as Jefferson did, among "degrees of genius."

And so we have the dichotomy: Jefferson regarded farmers as "the most valuable citizens"; Morrill looked upon the professions as "places of higher consideration." Morrill, in our time, as he must have been in his time, given the momentum of the industrial revolution, was a hard-headed realist sympathetic to the idea that we should "exalt the usefulness" of "those who must win their bread by labor." It would have been a hard sell to oppose it in that time because of the settlement pattern. Our government was responding to perceived local needs and problems.

Wendell's analysis is useful and necessary, but one wonders what would have been different if Jefferson's vision had been translated into the language of the land-grant college system. Could our rapacious nature dictated by the 3.45-Billion-Year Imperative have been confined or even slowed? Add to that the reality that this continent was settled and organized by white Judeo-Christian Europeans whose religious heritage included the biblical account of the Exodus, where Moses led the children of Israel through the desert into a Promised Land with the assurance that each would prune his own vineyard, sit under his own fig tree, and be his own priest. This Manifest Destiny idea of a "chosen people" must be tacked on to the 3.45-Billion-Year Imperative. It is part of who we are too. Our own version of Manifest Destiny featured a democratic distribution of land to mostly white citizens who could sit under their version of their own fig tree and, for some, be their own priest. There is nothing about "artificers" in the language of Moses and in the language of how to live upon the land coded in the some six hundred Mosaic laws. What that history is all about is being "Chosen People." What role did this idea of being Chosen play for us? Well, it has to be acknowledged, I think, for after the Civil War our government sent the military westward to kill Indians, which they managed to do with guns and diseases and through starvation, and by allowing white hunters to assist in killing off the bison. That job was finally finished off in the 1870s, and the first significant state of an extractive mining economy began, at least for hundreds of square miles within Jefferson's Louisiana Purchase. This is no trivial matter, but rather only a version of petri dish economics. The white hide hunters worked with great efficiency in a short period of time, gunning bison down so fast the barrels of their rifles often had to be cooled with water from their canteens, and when that ran out, with the urine of the marksmen. Bison meat did feed the railroad workers, but the number of bison killed for food was small compared to those shot for the hides.

Next came the miners of the bleaching bones, which they loaded in wagons by the hundreds to meet the new railroads. Now, with the wild bovine herbivore gone, came the miners of grass. The failure of the cattle industry due to the harsh winter of 1885 did set it back significantly, but only for a while. Next came the natural gas and oil miners, then the soil miners, and finally, now out over the Ogallala Aquifer, the miners of water.

With Manifest Destiny, backed by biblical scripture, backed by the 3.45-Billion-Year Imperative, how much can we blame the land-grant system and Morrill's inclusion of artificers to be educated?

During and after that bitter four years when, in Lincoln's words at Gettysburg, we were "testing whether that nation or any nation so conceived and so dedicated could long endure," people were wondering if there was such a reality as a benevolent God who let over half a million die. It is easy to imagine in their time—in spite of all—that the vast landscape now opening in the West was a deserved new beginning like a liberation. Something of an institutional nature happened; as Wendell said, those natives were surplus people, so we turned them into "redskins." At this moment we *de facto* institutionalized the idea of "surplus people." And as we marched into and squatted on the ancient lands of the native, little did we know that we had set a precedent about the acceptance of surplus people as the furnaces of industry grew in number to make people on the land increasingly unnecessary. More coal was being burned. Oil had been discovered. And because we were no longer a minor player in world affairs, but an emerging industrial giant, people economically forced off the land by the availability of fossil fuels were the new redskins. Is it a mere correlation that in the same period there were disputed elections and blunders by the military? The 1876 election was rigged, making Rutherford Hayes president over Samuel Tilden, Democrat from New York. The militia left the South. White southerners began to call the shots again. George Armstrong Custer in 1876, our centennial year, thought he needed a victory under his belt in order to return to the East and run for president. The defeat at the Little Big Horn in June dampened the enthusiasm for our centennial celebrated a few days after Custer died for our sins.

None of this seemed to slow us. Our exogenous version of petri dish economics accelerated. Pittsburgh steel was coming online, and Manhattan Island was being electrified. Death control accelerated. The germ theory of disease was verified, which contributed to the groundwork that ushered in antibiotics less than a half century later. Babies quit dying so often. The population, even without immigration, was poised to explode. But in the midst of it all, our institutions were becoming "more perfect" at acting on a perception. Professor Dan Luten at the University of California–Berkeley said it well. We came as a poor people to a *seemingly* empty land that was rich in resources. We built our institutions with that perception of reality. Our political institutions, our educational institutions, our economic institutions—all build on that perception of reality. In our time

Figure 1. Ina and Andrew Swan Ranch near White River, South Dakota

Figure 2. The Land Institute Sunshine Farm—a former ten-year project

Figure 3. Wendell Berry

Figure 4. Norman Rockwell's *The County Agent*, 1948

Figure 5. Soil ecologist Jerry Glover with four native prairie species (photo courtesy of Jim Richardson)

Figure 6. No-till and conventional till water discharge (photo courtesy of David Huggins)

Figure 7. The Harvesters, 1565, by Pieter Bruegel

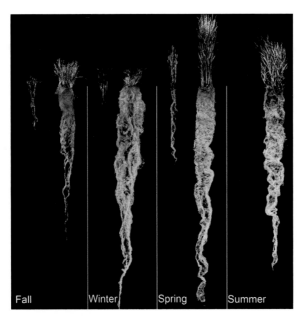

Figure 8. Annual vs. Perennial grain growth in tubes
(photo courtesy of Jim Richardson)

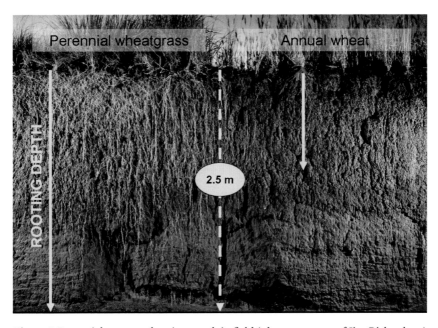

Figure 9. Perennial vs. annual grain growth in field (photo courtesy of Jim Richardson)

Figure 10. 1,000 intermediate wheatgrass perennial genotypes

Figure 11. Response to selection in intermediate wheatgrass

Figure 12. Hybrids between annual wheat and relatives showing regrowth

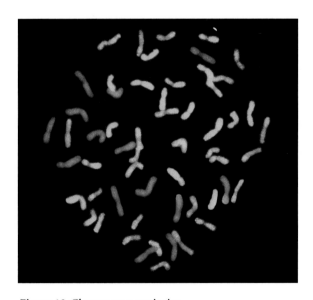

Figure 13. Chromosome painting

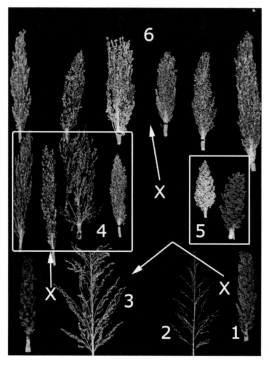

Figure 14. Steps toward a winter-hardy perennial sorghum

Figure 15. Perennial sorghum breeding nursery

Figure 16. Domesticating a nitrogen-fixing legume (Illinois bundle flower)

Figure 17. Seed shatter resistance, a necessity in domesticating a wild species (Illinois bundle flower)

Figure 18. Domestication of wild sunflower

Figure 19. Perennial and annual sunflower crosses

Figure 20. Perennializing upland rice in China

Figure 21. Perennial rhizomes in upland rice

Figure 22. Native prairie (foreground), perennial plots (background)

Figure 23. A slice of native prairie (photo courtesy of Jim Richardson)

Figure 24. The grasslands evolved to invite fire (photo courtesy of Jim Richardson)

Figure 25. The grasslands and the large herbivore (in this case, bison)

Figure 26. Selective grazers as a future management tool, the domestic bovine (photo courtesy of Jim Richardson)

Figure 27. Nonselective grazing as a future management tool

Figure 28. Nonselective grazing using solar-powered muscles (photo courtesy of *The Draft Horse Journal*)

Figure 29. Extra energy cost to speed hay drying (photo courtesy of *The Draft Horse Journal*)

Figure 30. Fuel for traction: draft animals and machines on farm

Figure 31. Illinois corn field, June 2008 (photo courtesy of Andy Larson)

Figure 32. Never-plowed prairie in the Kansas Flint Hills (photo courtesy of Jim Richardson)

we have become rich people in an increasingly poor land that is filling up, and the institutions don't hold.

To be fair, the momentum was there before the war to open the West, and the speed of our accomplishment can be illustrated by one small example of what men, horses, and mules could do in less than two years. Beginning in 1859, John Mullan of the U.S. Army commanded a crew to build a road from the Dalles on the Columbia through the pass at Coeur d'Alene to Fort Benton on the Missouri. It was 624 miles long and was finished in less than two years! The steamboat came to Fort Benton on the Missouri in 1860, the year it was finished. Here, through the northern Rockies, was a connection to the ocean, the Mississippi, and the Ohio. Now western emigrants could bring themselves and their supplies by water, and from there, on wheels across the northern mountains. Speed and efficiency characterized this restless nation—on the move.

High moments on the other side of 1862, and the emergence of free land and the land-grant system, would have canceled any careful introspection. There was Lincoln's assassination and our government's sending its armies westward to kill Indians and put the surviving natives on reservations. In 1869 the Union Pacific and Central Pacific lines met in Utah for the first coast-to-coast rail track, and as the first train rumbled over it for California, passing over the Green River in Wyoming, there was John Wesley Powell's men preparing to explore the last major river of the West. What need had we to ponder eventual limits and act on them?

How many even were aware of the ideals of Jefferson and his ideas on the requirements of good citizenship? What counted was the coal for the train and the mills and Pittsburgh steel and the oil out of Drake's well and the countless wells that followed. We were on the move. None of us wants constraints. Huck Finn had plans to light out for the territory. The local forces of Hannibal had imposed constraints on Huck he didn't like. But if Huck could disappear into the territory, he could be a boy forever, where there was freedom, infinite opportunity. I'm a direct beneficiary of that restlessness. The frontier was open to one of my own grandfathers, coming out of Kentucky to Kansas in the 1880s under an assumed name, running from something, and only after some years did he take his last name back and give my father his last name also. He was ever restless, eventually an alcoholic, later divorced from my grandmother in 1903. As he was dying from complications due to drink, he returned to her care to die. But the frontier had allowed him to escape whoever or whatever bothered him in the coherent community from which he had come. Like Huck Finn, he had lit out for the territory. He was not a "sticker" as described by Wallace Stegner, whose father was a "boomer", always on the move.

Where are we today? Unfortunately, the ecological limits are still obscure, and I sometimes think that we actually try to keep it that way. Lateral mobility made possible by a context with a large frontier and lots of resources still allows us a new start. This freedom means that we can evade having screwed up in a settled coherent context, and Wendell Berry's 1977 book, *The Unsettling of America*, has now taken on the dimension of a tragedy rather than a corrective that features a crisis of culture, economics, and ecology.

FROM 1862 TO 1935—SEVENTY-THREE YEARS— IS ONE PERSON'S LIFE

At the height of the Great Depression and the dust bowl in 1935, Franklin D. Roosevelt selected Dr. Hugh H. Bennett to be the founding chief of what was later called the Soil Conservation Service. It happened within the context of the most advanced thinking within a democratic society still on a new continent. "Big Hugh," as he was affectionately called, was described by the publishers of the *Farm Journal* as "one of the few immortals of agricultural history." Bennett was not satisfied to simply head another government bureau, for he saw his mission to be a "war without guns." The chief encouraged both individuals and groups to support soil conservation work and even went so far as to help establish a citizens' society devoted to similar goals for the American soils. To that end he was one of three people who met on the first of November in 1939 in Columbus, Ohio, to help establish an umbrella organization eventually called Friends of the Land. The other two were Charles Holzer, an MD, and Bryce C. Browning. Holzer had a clinic on the bank of the Ohio River at Gallipolis, Ohio, and had been motivated to do something about soil loss because of the connections he had made in his clinic between washed-out soil and sickly, washed-out people. He donated $1,000 to help launch the organization. Browning was active with the Muskingum Conservancy District.

Bennett was a holistic thinker and doer who felt strongly about saving the back forty. If the back forty were saved, society benefited. To Bennett the entire system of forests and plains, rivers and power, lakes and cities, was one. Bennett had traveled widely and strongly felt the need to make the interreliance of countryside and town clear to all.

People in various government agencies in Washington learned about this meeting from Big Hugh when he returned, and they joined him in endorsing the concept, though the bylaws of Friends of the Land prevented their service as officers.

Ultimately, most of the well-known names in conservation were involved in the organization: Liberty Hyde Bailey of Cornell; Gifford

Pinchot, the first head of the U.S. Forest Service; the cartoonist J. N. Ding Darling; ecologist Paul B. Sears; the novelist Louis Bromfield; and Aldo Leopold. All wanted the organization to be more than a pressure group. They wanted its members to develop a deep allegiance to sound soil conservation practices.

The organizers circulated a "Manifesto" in 1940 and sought to dispel any idea of threat to the other conservation organizations. Rather, they sought to "support, increase, and . . . unify all efforts for the conservation of soil, rain and all the living products, especially man." The word "ecology" is found but once in the founding manifesto, but most of the founders, whether farmers or professional people, were ecologists at heart. The organization got off to a good start, especially considering these were the days prior to World War II. Perhaps the success was partly due to the symbolically significant fact that Dr. Holzer was a physician and Dr. Bennett an engineer. Considerations of both the physical and biological world were united as one. The organizational meeting was held in Washington, DC, with some sixty men and women present.

It must have been an amazing meeting, judging by the group of people who attended. "There were artists present," recalled Paul B. Sears. A farmer-banker from Missouri had a thirteen-word summary of why he was there: "If you get an absorption of water, you prevent an erosion of soil."

The noted author Stuart Chase said at the meeting, "We are creatures of the earth and so are a part of all our prairies, mountains, rivers and clouds. Unless we feel this dependence we may know all the calculus and all the Talmud, but have not learned the first lesson of living on this earth."

This group of people, at once ordinary and extraordinary, was advocating a local response to a global problem. That they recognized it as such is clear from the author E. B. White's statement at the meeting: "Before you can become an internationalist you have first to be a naturalist and feel the ground under you making the whole circle."

At the third annual meeting, in 1942, the international theme surfaced again. Hugh Bennett was just back from visiting soil-conservation-trained people in Ecuador and Mexico. He commented that "when you get out on the land with people, and work with them and talk with them about the productivity of the soil, there is some sort of common denominator there. I think that our statesmen, our educators and all of our great men from the beginning of time, have missed that point.

"When you begin to talk and work with the fertility of the soil and the way it relates to the welfare of humanity, you are talking a common language. It brings people closer together. It will bring nations closer together."

At the same meeting the international need was further articulated by Dr. John Detwiler, president of the Canadian Conservation Society. "We begin to realize," he said, "that an overcrowding of people on a diminished soil base may impinge on the intellect, lead to physical and nervous disorders, and break forth ultimately in the hidden hunger that brings on wars. Perhaps when we organize conservation on an international basis we can avoid the hidden hunger which brings on wars." In 1945, Detwiler sent a communication to the organ of the organization, *The Land Quarterly*. He said, "To preach conservation at such a time, when all our resources, national and otherwise, are being sacrificed in unprecedented measure, might seem to some anomalous, even ironical. . . . But we firmly believe, and now are more acutely aware than ever that conservation is basically related to the peace of the world and the future of the race." The journal carried an "Other Lands" section as further indication of the group's awareness of the connection between prosperous soil and peaceful people.

From the outset, the leaders, many of them government workers, recognized that *The Land Quarterly* "should be written from the ground up and not from Washington down." Symbolically, they moved the central office from Washington to Ohio in 1941. It was about this time that Louis Bromfield of Malabar Farm joined the organization and devoted much time, money, and his own literary skills to the organization and the cause.

Let us consider for a moment the organization in *government* these people were trying to support. Bennett had quickly assembled the most able engineers, agronomists, nurserymen, biologists, foresters, soil surveyors, economists, accountants, clerks, stenographers, and technicians of many backgrounds into the service of saving the soil. There was an anxious urgency on the part of nearly everyone involved. The Soil Conservation Service was, as Wellington Brink put it, "born with pride and loyalty and a sense of high destiny—an inner element that was to persist and spread and animate the organization and weld it together with a spirit altogether unique in modern government." Because of the high caliber of the people employed, the SCS gained a good reputation fast.

The SCS was an organization based on four steps: science, farmer participation, publicity, and congressional relations. Bennett explained these four steps in his characteristic North Carolina accent many times. "By science," he would say, "we tried to imitate nature as much as we could. We abided by the following basic physical facts, (1) land varies greatly from place to place, due to differences in soil, slope, climate and vegetative adaptability; (2) land must be treated according to its natural capability and its condition as the result of the way man has used it; (3) slope, soil, and climate largely determine what is suitable protection in all situations. . . . Above all . . . we tried to imitate nature."

The farmer-participation step was extremely successful in helping farmers become interested and involved. SCS men would walk over a farm, field by field, with the owner. There were tens of thousands who came to the famous soil conservation demonstration meetings. The entire approach was touted as a most democratic and practical approach to encouraging farmers to safeguard the natural resource. When Bennett retired, the program involved nearly five million farms. And here is a point we need to come back to later on. The per-acre crop yields began to rise steeply between 1935 and 1950. Because there was a correlation between sharp crop production increases and the increasingly widespread soil conservation practices, it appeared that soil conservation was the cause. But, again, more on this later.

Bennett's third approach, publicity, seems to have been fully exploited. In his own words: "We employed capable writers, trained journalists who knew what to write and where to get articles published. They were not agronomists or agricultural engineers trying to be reporters or advertising men. They were journalists who knew good pictures, good news angles, good feature possibilities when they saw them. And I let them go at it, in their way, which happened to be the newspaper and radio way. The country papers picked up our . . . material by the volume and it got back to congressmen. . . . Clippings piled up, requests for more information poured in. . . . We considered the press and radio a very important part of our job." USDA records will show that Bennett was right.

In the last step, congressional relations, Bennett's charming and colorful personality ensured success. He kept the congressmen informed of the conservation effort in their home districts and talked to them about plans for the future. Bennett became a legend on the Hill. "When I appeared before a committee," he explained, "I never talked about correlations or replicas. But I did spread out a thick bath towel one day on a table before a committee, tipped the table back a bit, and poured a half pitcher of water on that towel. The towel absorbed most of the water, cutting its flow from the table to the rug.

"I then lifted the towel and poured the rest of the pitcher on the smooth table top, watching it wash over the edge onto the rug. I didn't say anything right away—just stood there looking at the mess on the floor. Then I looked up at the committee and explained the towel represented well-covered, well-managed land that could absorb heavy, washing rains. And that the smooth table top represented bare eroded land, with poor cover and management on it. They seemed to understand, because we got our appropriation."

This was the organization and the chief that the Friends of the Land sought to support. The society was itself free of government, unpretentious, informal, friendly, devoted to "rational evangelism" with high-level discussions and lots of adult education. Their mission was education, and they believed in the democratic way, sometimes (one is tempted to think) to a fault. One member once wrote that "no lasting reform can possibly be accomplished by compulsion, by bribery, or by regimentation. Information, knowledge, *education* that extends to all callings and all age groups of the people—that is the only democratic way."

In a review of their works and aims, their editor commented, "It was not our purpose to conduct lobbies, either in Washington or at the state capitols, but rather to attack by meetings, tours, institutes, publications, press dispatches, articles, books, movies, radio and all educational media a prevailing public ignorance and inertia."

"This is slow but it has the advantage of being a frontal assault," a friend wrote in the *Baltimore Sun*. "If carried through, [it] will be final. For if the American people were actually informed and alert on conservation, then programs of legislation would be merely matters of detail."

I suspect that never has such a large group of people seen so many links in the cause-effect chains of nature. On the cover of several issues of *The Land Quarterly,* a well-written and artistic publication, is the statement, "A Society for the Conservation of Soil, Rain and Man." Aldo Leopold, who wrote articles for the journal, reminded readers in his rich prose of the connections between people and land. It was he who wrote in *A Sand County Almanac:* "Health is the capacity of the land for self-renewal. Conservation is our effort to understand and preserve this capacity."

The organization was phased out in the late 1950s, but their publication revealed that it was sinking before then. How could such an organization go under, an organization that held hundreds of meetings, often at summer camps, at teachers' institutes, and at metropolitan centers? They traveled to the hinterland and held public forums. They were not cultists or fanatics. They simply wanted to explore the facts and pass on the information. It was typically American, like the Farm Bureau or the Chamber of Commerce. It was not nearly as sexist as many organizations were at that time. In 1949, the fifth president, upon taking office, said, "The care of the earth for all the years to come is not exclusively a stag affair."

The organization probably went under because of the nature of the 1950s, which included a great acceleration in the use of fossil fuel. It was an era of great complacency, too, in America. Technology was solving our problems. Everything was going to be bigger and better, and the opportunities were limitless. It was not a time when many were inclined to listen to the dire warnings of a few who happened to be reading different signs of the time.

Stand back for a moment. If the year were AD 1930, one might conclude that Job and Isaiah and all the other Hebrew prophets were early voices of a minority that had extremely sharp vision. Patrick Henry and George Washington, Henry Thoreau and Ralph Emerson, John Muir and George Perkins Marsh were only the perceptive and eloquent observers on a continent that was then regarded as yet young and endless in its resources. One might contend that all these greats constituted the important historical stream of pioneers upon whose shoulders a movement finally stands.

By AD 1940, with the Soil Conservation Service and supporting organizations, we might be comfortable that it was all finally coming together. Government, industry, and private organizations were winning the war against soil loss.

By AD 1950 we who are given to examining correlations might have nodded our heads knowingly and restfully as we noted that gullies *were* arrested and *production had increased 20, 30, 40, 60, and even 100 percent*. Other governments would *now follow* our model.

We should not forget that there was a dramatic yield increase between 1935 and 1950. Much of the per-acre increase was due to retiring marginal land from production. For example, if we have two acres of corn, one that produces 120 bushels and another that produces 60, the average would be 90 bushels per acre. By retiring the low-producing acre, we immediately experience a 33 percent increase in per-acre production.

For purposes involving federal budget allocations alone, one might expect SCS officials to take as much credit as possible for the increases in crop production even though the farmers who were paying for the fertilizers and pesticides might have thought otherwise. To the extent SCS scientists were taking credit for production increases, they were also ignoring one of the oldest warnings of science: scientists should be careful in their talk about cause and effect and speak first and foremost of correlations. For, if we are to take seriously the impressive studies from the early 1970s to 1976, annual soil loss was greater by at least 25 percent than in the dust bowl years when the SCS was begun.

The SCS certainly did help the country in setting aside marginal land. Who wants to imagine what would have happened to our soils without the SCS? We cannot assume, however, that SCS-supported activity is the primary cause for higher production, except in a few places. We should look to the increase in fertilizer, pesticides, improved varieties, and irrigation. This is a very important point, for as I have already stressed, our fossil-fuel-based chemotherapy prevented us from seeing the wasting away of our soils and gave us an unrealistic view of actual soil health.

The painful conclusion we must live with is that both governmental and citizens' organizations have failed to prevent soil loss.

The 1977 Soil and Water Resources Conservation Act (RCA) was cited by the USDA at the time as the most important legislation since the first soil and water conservation measures were implemented in the 1930s.

The USDA proposed a major overhaul of programs throughout its thirty-four separate administrative offices in order to meet anticipated demands on the nation's natural resources over the next fifty years.

The implementation of the legislation began with public hearings in 1978 to identify problems in the conservation effort. I went to the one held in our district. It was poorly attended; according to one official at the meeting, only one or two people represented some of the districts. The meeting was very disappointing, and I got the feeling one experiences at such meetings that it was being conducted more out of simple compliance than out of any earnest effort to elicit spontaneous elaboration about what was wrong with the conservation program. I suspect that most meetings, all across the country, went this way. Nevertheless, according to the summary statement of all these meetings, the most often repeated complaint was that farmers needed more technical help from the Soil Conservation Service in order to both plan and implement conservation programs.

Following the nationwide public hearings in 1978, seven resource areas were identified, and objectives for each of the seven were established. Next the USDA proposed seven conservation strategies to meet the proposed objectives. At the hearings, farmers, ranchers, and interested citizens were asked to consider the proposed policies, which ranged from totally voluntary compliance to complete regulation. One extreme regulatory measure, if it were approved by the president and Congress, would have required landowners to apply acceptable conservation measures or be denied any assistance from USDA farm programs.

The potential effectiveness of any new program in the conservation of soil and water will depend not only on the new rules handed down by government, but also on the *owners* of the land. New ownership patterns were emerging rapidly by 1978, as they had been for the previous twenty years. Over half of all agricultural land had changed owners since 1960, and at the time, 2–3 percent of all agricultural land was changing ownership each year, according to the RCA review draft. Land became increasingly an item for speculation, and absentee ownership became widespread.

One reason the RCA effort was hampered is that with fewer families on the land there were fewer people to love the land in any way close to the way it deserves and needs to be loved. RCA administrators anticipated this reality and expressed their concern in their 1980 review draft.

The core of this chapter was published in *New Roots for Agriculture* in 1980. I expressed my hope that the RCA program would raise the level of consciousness about the need to conserve. More and more, prime farmland was being taken out of production by urbanization and industrialization, which forced a shift to more intensive agriculture on nonprime acres. More energy-intensive fertilizer practices were used and production soared.

In the mid-'80s the Conservation Reserve Program (CRP) was initiated, which at the outset took 36.4 million acres out of production over a ten-year period. This cost the taxpayer an average of $500 an acre for a total cost of about $18.2 billion. And now the program is being effectively abandoned.

The United States, relative to its level of consumption, is resource-poor apart from forests and soils. The temptation for any administration is to spend these resources to minimize the balance-of-trade deficit. In 1977 the cropland base was 413 million acres. According to the National Resource Inventories, nearly one-fourth of this acreage, 97 million acres, has eroded in excess of five tons per acre per year on average. Since the minimum-till and no-till techniques were developed, made possible by the use of herbicides, we effectively were acknowledging that we have to poison our soils to save them. But even accepting the need for herbicides, soils in the Upper Midwest are cold and wet in the spring, and with stalks reflecting light, the seeds rot before they germinate. Soil erosion continues.

In summary, in spite of all our efforts at conservation of our soils, humans are now the primary earthmoving agents. During the ice age, glaciers deposited an average of about ten billion tons of till in moraines and outwash fans every year. Agriculture today contributes as much dis-placement as the glaciers of the Pleistocene, and agriculture is not alone. The total movement of earth by humans now is estimated to be around forty to forty-five billion tons per year. We are the most important agent currently shaping the surface of the earth, according to geologist Roger Hooke. The glaciers were giving us fertility by pulverizing the rocks, releasing essential minerals available for life. We agriculturists, in the form of soil erosion, send those valuable minerals toward a watery grave. Nearly 40 percent of the soils of the world are now seriously degraded. Globally, nearly one-third of the land devoted to farming has been lost to erosion since 1960, and continues to be lost at a rate of some twenty-five million acres per year.

The problem of agriculture remains, but it can now be solved.

CHAPTER SEVEN

CHEMICALS ON THE LANDSCAPE

The elements that make up living creatures are those that are common on the terrestrial surface or within easy reach of roots. In our oceans they are made available through tidal or wave action. There are only twenty-some such elements—depending on the organism—that are at once common and necessary. The periodic chart lists 114 elements. A convenient memory aid for countless chemistry students for some of those twenty-some was CHOPKINS CaFeMg. Carbon, hydrogen, oxygen, and nitrogen are all atmospheric elements that circulate throughout the global commons. The common land-based elements are phosphorus, potassium, iodine, sulphur, calcium, iron, and manganese. These latter elements are more at home on the back forty than floating in the atmosphere. All of organismic life is the consequence of complex chemistry with these few common elements. Humans as chemists, however, have used nearly *all* of the elements represented on the periodic chart to create what the advertisers have called "better living through chemistry." Among these creations we should list thousands of synthetic poisons and toxins.

But there is another difference between the chemistry for organisms and the work of human chemists. Organisms feature complex reactants with a few elements, whereas chemists feature simple reactants with many elements. Professor Terry Collins at Carnegie Mellon University, in charge of the Green Chemistry program, begins with this distinction between nature's chemical world for organisms and that of the human chemist. He uses this as the first consideration in determining what is safe and what is not to be introduced into the world. For example, an electric eel, made of complex reagents, is a complex creature we might eat with impunity. A lead acid battery, by contrast, is a simple creation, but we had better not eat it. While nature creates complexly and safely, we create simply and dangerously.

Professor Collins concludes that the release of humanly made compounds into the environment should occur only at rates that are compatible with natural biological processes. Persistent synthetic (not naturally occurring) substances should not be released at all. Here we see a concrete example of how cultural development is not in step with scientific/technological advancement.

By ignoring the implications of this Green Chemistry, we continue to poison the planet and ourselves. The implications with agriculture are serious. At best, 1 percent of applied pesticides reach their intended targets. Since 1950, insecticide usage in the United States has increased from 15 million pounds to more than 125 million pounds. Even so, it has been

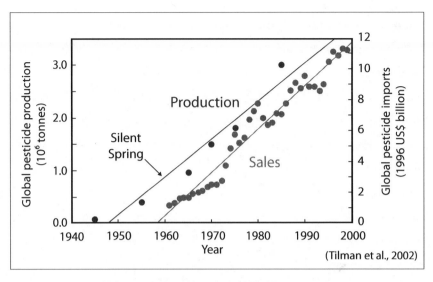

Figure 16. Post–World War II production and sales of pesticides increased more or less parallel with increase in use of commercial fertilizer (see *fig.* 14, page 84)

a losing battle. Over this same period, crops lost to insects nearly doubled from 7 percent of total harvest to 13 percent and, in that time, numerous studies have verified the suspected link between pesticides and diseases. This has been no small task, because direct links are often impossible to establish. Experimentation employing direct dosages is usually regarded as unethical. Consequently we are left with epidemiological evidence. A summary of cancer risks among farmers states that "significant excesses occurred for Hodgkin's disease, multiple myeloma, leukemia, skin melanomas, and cancers of the lip, stomach, and prostate" due to pest control chemicals. Another study posits that the herbicide 2,4-D has been associated with a two- to eightfold increase in non-Hodgkin's lymphoma in agricultural regions. The study of farm chemicals and their evident role in disrupting the human endocrine system is a fast-growing field. One report reveals that several pesticides can reduce the immune system's ability to deal with infectious agents. Formerly, acute poisonings and cancer risks dominated risk assessment. No more. Now direct evidence from clinical and epidemiological studies shows that people exposed to pesticides experience alteration of their immune system structure and function.

The neo-agrarian will take seriously the implications of Green Chemistry, just as he or she will believe that soil is as much of a nonrenewable resource as oil. Human invention will be brought into line with nature's own creative patterns.

Chapter Eight

THE LOSS OF CULTURAL CAPACITY

The July 24, 1948, cover of *The Saturday Evening Post* shows a painting captioned *County Agent* by Norman Rockwell. The painting's permanent home is in the lobby of the Nebraska Center for Continuing Education in Lincoln. It's so popular that thousands of prints can be found in local extension offices of the Department of Agriculture throughout the United States (see color insert *fig.* 4).

Rockwell used real people as models, and in this case it was a real county agent named Herald K. Rippey, whose charge was Jay County, Indiana, and the Steeds, an active 4-H family of Portland, Indiana. A Nebraska 4-H Development Foundation pamphlet written by Clarice Orr provides an interesting history of the painting:

Artist Rockwell followed County Agent Herald K. Rippey around Jay County, Indiana, and, according to the story, ended up "worn to a nubbin', but chock-full of farm cooking, tips on how to cull chickens and test soil, and warm admiration for his subject."

Clint Hoover, director of the Nebraska Center Hotel, spotted Mr. and Mrs. Rippey, one summer day, standing in the lobby in front of the painting. En route home from a West Coast vacation, they stopped to see the painting in its permanent home. Rippey reminisced about 4-H and his brush with fame with Norman Rockwell. . . .

When Mr. and Mrs. Rockwell arrived in Indiana, he was well pleased with the plans for the setting except for the barn—it was white and he wanted a red one. . . .

Although he planned for a spring scene on the canvas, preparations were actually done in the fall. Faithful to detail, Rockwell changed the boy's winter cap to the proper spring attire. And after Rockwell's farmer neighbor reminded him that fall calves have heavier coats than spring calves, Rockwell repainted the calf.

Rockwell painted what was writ large in American agriculture. The father, mother, and the grandfather all stand aside for guidance by the visiting government employee. The horse is confined to the barn. Expertise and youth are central. Tradition and experience are peripheral. The horse, the cats, the chickens, the dog, all useful pets on a 1948 farm. The girl with the sewing project was learning to contribute to the family economy. She was developing her cultural capacity.

This painting captures an important transition in the upward and outward direction of agricultural science and technology. Disappearing are the days of reliance on human and draft animal muscle power. The knowledge in the heads of the grandfather, father, and mother is knowledge keyed to the necessity of planting, harvesting, and the storage of contemporary sunlight from the land. By the 1940s, as Wendell Berry has said, "animal husbandry is becoming animal science." The three adults are both trained and educated to be husbands of our resources. The young are being told by the government expert what science says about improved farming.

Time and again we're told that the current scale of industrial agriculture was an inevitability, or that economic determinism was at work. Efficiency—how few farmers are necessary to produce our food—is regarded as a virtue. A sign on Interstate 70 near the Manhattan, Kansas, exit—the exit that also leads to Kansas State University, our local land-grant university—brags to the passerby that one Kansas farmer feeds 128 people plus you. The number has been climbing over the years.

What this means is that the sufficiency of people, the higher eyes-to-acres ratio, is being supplanted by a sufficiency of capital. And the sufficiency of capital is almost totally dependent on the increased availability and use of fossil carbon. Rockwell's "expert" instructs the farm children on the fossil-fuel-derived farming system, replacing the older members' wisdom, hard-won as farmers schooled by the vicissitudes of nature.

For years I wondered, What became of the Steeds? Mom and Dad and Grandpa must be gone, but where are the kids? I asked the telephone operator in Portland, Indiana, for the number of any Rippey in the phone book. I dialed one of the offered numbers, and a Mr. Rippey answered, a relative of the late county agent. From him I learned that the father in the painting, who stands off to the left in the shadows with the cat on his shoulder, farmed until he was killed by a registered bull on his farm. Mr. Rippey gave me the name of the dutiful 4-H girl holding the calf, the one with her 4-H record book displayed, and I called her. She and her sister, the one with the sewing project, left the farm to marry. They never farmed. Their brother, the one with the poultry project, had farmed until he died in 1988. I asked if any of the children of the kids in the painting were farming. The answer was no.

Such stories are commonplace in studying the decline of American agrarianism. You also hear about the desire for urban life's stimulation, for upward mobility, which lateral mobility makes possible, or the wish for ease and the understandable push to make a farm more productive.

What the painting does not show is people leaving the land. World War II had been over almost exactly three years when the painting appeared. The science of agriculture was growing. I was twelve when that picture appeared. In another two years I would enter a rural high school and learn the Future Farmers of America (FFA) creed: "I believe in the future of farming with a faith born not of words but of deeds— achievements won by the present and past generations of farmers; in the promise of better days through better ways." Who can argue with that? But the creed continues, "even as the things we now enjoy have come up to us from the struggles of former years." Those struggles include coping with a limited supply of muscle power, which in turn was tied to the cycles of nature, growing plants, and feeding livestock. Those limitations created a kind of frugality. And lots of hard work! But the creed shows that it isn't all struggle: "I believe that to live and work on a good farm is pleasant as well as challenging; and hold an inborn fondness for those associations which, even in hours of discouragement, I can not deny." The creed has changed twice since to include the reality of industrial agriculture and the decline of rural America.

WHERE ARE WE TODAY?

We live in a time in which the ecological limits are still obscure. Mobility has moved us beyond the local bank, where membership in the community formerly forced us to cope with the limits of the community. Lateral mobility, made possible by a large frontier and lots of resources, has always allowed a new start. A certain freedom means that we can evade having screwed up in a settled, coherent context.

I think about what disappeared after World War II. With our cultural capacity intact, we could then accumulate abundance or cope during a slack period. What disappeared is our cultural capacity; to adjust cultural information is complex, hard-won, but fragile. It is accumulated and practiced day by day.

And now what we have is a lien against the future. Our sources, our resources (oil, natural gas, soil, and metals) are being used up.

CHAPTER NINE

GLOBAL WARMING: HOW A LITTLE CHANGE CAN MEAN A LOT

There once was an evolutionary biologist in the Soviet Union who found himself caught between the political views of T. D. Lysenko, a man who believed that the environment could directly alter heredity, and the views of traditional genetics, which featured narrow determinism. This man was named Ivan Ivanovich Schmalhausen. One of his insights has come to be known as Schmalhausen's Law. Harvard ecologist Richard Levins described what Schmalhausen had observed as follows: "Organisms near the boundary of their environmental tolerance or geographic range are very sensitive to local conditions. Trees that grow on any soil type in most of Eurasia are limited to limestone soils near their northern boundary. Soil differences that did not matter in Central Russia were a matter of survival at the edge of the tundra."

It seems intuitive enough that, if we were to trace the boundary of the range of a species, we would be likely to discover that such organisms would display a greater range of variation because they are stressed near their boundary of tolerance. As Professor Levins explains, the homeostatic mechanisms, evolved to function well when conditions are normal for that organism, are less adapted under the unusual or extreme conditions.

Several years ago Professor Levins was involved in an experiment with corn in the University of Wisconsin Biotron. The computerized controls allowed the experimenters to select the precise environment they wanted. Many local corn collections, varieties, and hybrid lines were grown at normal conditions. He and the students also grew similar genetic lines near the boundaries of their tolerance, with temperatures that varied from too hot to too cold. They measured growth twice a week. At the normal conditions for corn, the varieties grew uniformly, and even the differences among varieties were moderate. But under the extreme conditions that the corn hadn't met before, each variety was more variable, and differences among varieties also increased. When faced with dry, cold days and hot, wet nights, only two varieties survived. One was identified as being drought-proof in the Midwest, while the other was an accession from Mexico. Neither had any outstanding characteristics in the normal conditions.

Levins explained the differences within the varieties as follows: First off, even a controlled environment chamber will have some fluctuations; small accidents in the internal flux of molecules as development proceeds will yield differences. And there will always be some genetic differences even in presumably homogeneous varieties. What the extreme conditions did was amplify the effects of these subtle events. The differences among varieties were caused by the same factors, but also by more genetic differences that didn't matter much under normal conditions but were revealed near the boundary of tolerance. (The external temperature and light flux and the variations in DNA were "external conditions" from the perspective of plant growth.)

The story gets increasingly scary when Levins describes how the same principle applies to insect development. Entomologists know that an insect completes its development when it has accumulated a certain number of degree days of heat about some threshold. Professor Levins walks us through. Suppose it needs 100 days above 14°C and that the actual temperature is 20°C. Then its development time would be $100/(20 - 14) =$ 16.7 days. If the temperature changes to 19°, development time would be 20 days. One degree change results in a 12 percent change in development time. But if the actual temperature is 16°, it would need 50 days. A change of 1° here halves the development time. Near the boundary, a 1° change had a lot more significance. Suppose now that a second species had a temperature threshold of 15°. At 16° it would need 100 days, but at 15° it wouldn't develop at all. Here a small genetic difference or a tiny difference in temperature has a big effect near the boundary, while farther away the difference is small.

On April 24, 2008, National Public Radio's *Morning Edition* broadcast two stories related to global warming. The first story was from British Columbia. It reported that ninety million acres of conifer forest had been destroyed by pine beetles and the fungus they transmit. This forest was supposed to help make Canada carbon-neutral. Any gains from other carbon-control measures in Canada will be negated as the forest decays or burns. The story continued by noting the widespread budworm infestations on spruce in eastern Canada and widespread pine moth infestations on pine in boreal Siberia. The reporter and his sources managed largely to blame the insects, though it was clear from the report that the problem is that earlier springs and later falls mean more generations of insects and warmer winters mean fewer winter die-offs. Systems theory people would call these phenomena biotic positive feedback. From a human point of view, the word *positive* here is not a plus.

NPR also reported that Siberia was experiencing numerous and extensive forest fires, over nine hundred, due to a serious drought. Enormous amounts of carbon are being released suddenly. Carbon absorbers are being turned into carbon producers.

The consequences of climate change, carbon sequestration, and carbon trading to offset the consequences of burning fossil carbon—well, how many carbon traders on Wall Street should we suppose have ever heard of Schmalhausen's Law, degree days, insect development, and the need to check with an ecologist or two before the economic grand proposals are launched?

Levins's message is that at the extreme edges of tolerance, almost anything can destabilize a system. There will always be some precipitating factor. If we remove one, then another will produce the outcome. What provokes the final change is less important than the vulnerability itself.

But if all variables interact with each other, if choices are links in a network, both determined and determining of outcomes, where do we locate causation? Richard Levins again:

> There are two ways of looking at causation in complex networks. The first is to ask, what external factors, touching the network at a particular place, are responsible for a set of observations. For instance consider a community of species in the plankton of an estuary. The variables of concern may be several kinds of minerals, perhaps 100 species of algae, crustaceans that feed on the algae, crustaceans that feed on those crustaceans, fish, fishing boats and so on. If a change is induced by an increase of minerals running off the land, then the algae increase, the crustaceans have more food and so on up

the trophic hierarchy. They all increase together. But if the initial change is in day length, acting on the algae, a longer day increases algae directly by increasing photosynthesis and thus decreases minerals since they are taken up by more consumers. The other levels will increase along with the algae. But if the initial change is an increase in fishing, alternate levels in the trophic hierarchy decrease and increase. Thus we can explain the observed pattern in terms of the point of entry to the system of the driving force. In one sense the "cause" of the observed changes is the increase in fishing.

But at a deeper level, the "cause" is the whole network. The outcomes of particular events are what they are because of the structure of the food web and the negative feedbacks between consumers and consumed. This ecological law even has moral implications when it comes to social justice since, as Levins explains,

> It is the marginality itself which makes people vulnerable to trivial differences of experience, skill, energy, acuity or inclination, almost making perfection a necessity and loading choices with grave consequences. In a marginal situation—poverty, oppression, inequality, stress—the human system may become injured in many different ways. These pathways cross back and forth across the boundaries we impose to distinguish social, behavioral, economic, biological, psychological and other processes. But tracing the precise pathway by which poverty does its damage cannot be allowed to detract from the more general conclusion that social deprivation is bad for our health."

PART III

Reversing the Damage

CHAPTER TEN

CONSULTING
THE GENIUS

CRACKING THE OLD CHESTNUTS

A problem that is hard to crack is sometimes referred to as an "old chestnut." The old chestnut for agriculture has two parts: (1) How do we obtain an adequate, if not bountiful harvest? and (2) How do we ensure that future adequate or bountiful harvests have not been compromised during production? The first question requires farmers to explore ways to maximize sun-sponsored fertility through ample supplies of moisture and protect the crop from insects, pathogens, and weeds. The second question causes us to look for ways to minimize soil erosion and avoid chemical contamination of our land and water and, in our time especially, reduce the necessity for fossil fuel inputs.

The challenge and the necessity grow by the day. For example, since the famous conference and three-volume report in 1956 on *Man's Role in Changing the Face of the Earth*, the world population has more than doubled. In our time, a half century later, according to a recent study by scientists Navin Ramankutty, et al., nearly 40 percent of the ice-free land surface of the planet is now devoted to supplying food for humans. To gain access to this new land, on a global basis, most of this new acreage has come from land never cultivated, which means that nature's ecosystem structures are being dismantled and the functions accompanying them have been compromised, some severely. Increase in global warming is but one consequence. It is more than a source problem. It is also a sink problem. Just clearing of forests in the tropics for crops or grazing is variously estimated to account for 12 to 26 percent of the total carbon dioxide emissions to the atmosphere. Some 30 percent of all surface water is being drawn to irrigate land to produce our food. And now with the added demand for biofuels as the fossil liquid fuels run short, the water requirements for growing corn skyrocket. Charles A. Washburn said it well with respect to corn:

> Corn typically needs about 24 inches of water, either precipitation or applied, during the growing season; if that yields 220 bushels/acre corn, the "depth" of the EtOH produced is 0.022". If we assume just 12" of water was applied to supplement rain, the ratio of water/EtOH is 553:1. When you see a railroad car filled with ethanol go by, think of the five trains of water, each with 110 cars, hauling the water necessary to produce one tank car of ethanol.

Looked at superficially, it may appear that we are doing all right. On a global scale more food than ever in history is being grown. But there is the accepted dualism: agriculture unintentionally but tragically worsens the global ecological problems and is heavily reliant on nonrenewable resources. These two faces of agriculture—productivity and destructiveness—do not arise from the conscious decisions of people who sell fuel, fertilizer, or pesticide, or of farmers, government officials, or grocery shoppers. As I have discussed, productivity and some form of destructiveness have existed in the way humans have practiced agriculture for over ten thousand years. Recently, chemicals and other nonrenewable resources have only exacerbated the problem, most especially when considering annual monocultures for grain.

CONSULTING THE GENIUS OF THE PLACE IN THE TROPICS

Ecologist Jack Ewel's experience of many years in Costa Rica allowed him to see firsthand that the traditional farming system featuring slash-and-burn had a rotation length increasingly shortened due to increased food demand. With shorter cycles fewer nutrients were accumulating in the standing forest, and so when cleared, subsequent crops had less available from the ash bed. Yields were in decline, and soil erosion and nutrient leaching increased. What had worked for centuries, probably millennia, as a cycle of restoration was now a cycle of degradation. And it was not just in Costa Rica where this pattern existed; tropical forests around the world were under siege.

Having observed all of this with an ecologist's eye, in the mid-1970s Jack Ewel designed what has become an increasingly famous experiment. The outcome of that study should help set the stage to explain our work at The Land Institute and the primary proposal for an alternative approach to agriculture on a global basis.

Jack Ewel and his fellow scientists, since the mid-1970s, had studied successional vegetation there as a model for the design of sustainable agriculture in the tropics. They were looking for standards for designing farming schemes. They reasoned that if they could channel some of that natural productivity into food, fuel, fiber, fodder, or cash, they might be able to develop economically viable alternatives to the destructive slash-and-burn methods settlers use.

They had some basic science questions in mind as well. Unlike most agricultural crops, successional vegetation is rarely devastated by pests. They wanted to unravel the mechanisms that endow these natural communities with resistance and resilience to pest attack.

In addition to these two objectives, they had a third of primary interest here. Successional vegetation seemed to afford excellent soil protection. Most kinds of cropping schemes in the high-rainfall tropics are all too often accompanied by nutrient leaching and soil erosion. Addiction to expensive chemical fertilizers is one consequence, but even worse is the semipermanent impoverishment of the soil.

After studying nature's diverse, fast-growing creation, they set out to build and study a parallel community, one that contained as many species of plants and in the same proportions of life-forms—herbs, shrubs, trees, vines, palms, and so forth—as the natural community.

Some plants in the mimic were cultivars, some wild, but none native to the site. But the mimic species had to resemble some plant that occurred naturally in the successional vegetation, so the Ewel team substituted vine for vine, herb for herb, tree for tree, shrub for shrub. These ecologists were careful to select plants that could not have found their way to the site without the help of the team. The substitute plants had to be both structural analogs and aliens. Both the natural and the mimic communities were species-rich and structurally complex.

These scientists spent a year perfecting their horticultural skills. And what did they find? The composition of the natural community changed rapidly, and in that first year it beat the mimic in biomass production soundly: 1,800 grams versus 1,150 grams per square meter. By the fifth year their efforts at imitation paid off. The productivity of the mimic came to more than 90 percent of the model. That's the production side of the story.

As for pests, the two communities lost leaf tissue to insects at about the same rate, but in the face of a pest outbreak, the mimic fared worse than the natural vegetation model. The natural successional vegetation was only slightly more complex than the mimic.

The take-home message is that early on, nature's tropical rain forest "genius" pumped much more water *back into the atmosphere than did the imitation.* In the rain forest, water is a nemesis to fertility, and for a year or two the mimic suffered from having 23 percent more water moving *downward* than did the successional vegetation, which meant more essential nutrients were leached—three grams of nitrate-nitrogen per square meter, for example. And that nitrate washed out along with a gram each of calcium, magnesium, and potassium.

Within a couple of years, however, leaching rates in the two communities were indistinguishable. This mimic was as water- and nutrient-tight as the model. This is good news for those of us who want to look to nature's ecosystems for a new agriculture. Good news indeed, for an adjacent plot of bare soil lost water 50 percent faster than plots that were vegetated. Nutrient losses from bare soil were even more dramatic. Here losses of nitrogen and cations were ten times faster than either the mimic or model. I like the way Jack put it: "[Since] bare soil is the nemesis of fertility in the humid tropics, without the return of water to the atmosphere through transpiration, and in the absence of nutrient uptake by plants, leaching is rampant and soil impoverishment is inevitable."

This team verified the fact that nature's way worked pretty well, that local nature was the genius. They found that if they mimicked the structure, they were usually, but not always, in Jack's words, "granted the functions: high productivity, responsiveness to pests, and good protection of the soil."

They got the kinds of basic understanding they sought, but they were well aware they had no technological package they could hand over to tropical farmers. Such diversity would be an agricultural nightmare. However, they concluded, the immense richness of species they worked with was likely unnecessary both agronomically and ecologically. Simpler communities might provide many of the same benefits, provided the plants they contain are capable of high productivity, either resistant or resilient to pest attack, and able to use the soil thoroughly and continuously.

WHY PERENNIALS?

For the first time in over ten thousand years of agricultural time we can now imagine a sustainable agriculture. A convenient beginning point is to consider that essentially all of nature's land-based ecosystems feature perennials. This has to do with the fact that only twenty-some elements represented on the periodic chart in our science classrooms go into all of earth's organisms. And importantly, only four of these building blocks circulate in the atmospheric commons. The rest are in the soil, and they are all water-soluble. Managing these elements and water is accomplished by a diversity of roots below the surface. This management scheme occurs in millimeters and minutes. Alpine meadows, tropical rain forests, desert scrub, deciduous forests, and native prairies all feature perennial mixtures.

It is with this last reality that a crucial part of our story begins: the elegant micromanagement of these essential elements has eluded farmers who work with annual monocultures. Annual monocultures on the landscape are not there year-round. Their extent is usually limited, depending on the species. In such a manner we built an agriculture that was at once simple and simplifying, disrupting countless subtle, ancient processes that had been reliable over millions of years.

We can't go back to the crossroads where our ancestors took that wrong turn, or to a golden age of folk agriculture that never existed. But we can now envision an agriculture in which we bring the ecological processes embodied within wild biodiversity to the farm, rather than forcing agriculture to relentlessly chip away wild ecosystems.

DIVERSE PERENNIAL SOLUTIONS

Since 1976, The Land Institute has been working on the idea that humans can make conservation a consequence of production—in any region on the planet—and in doing so, we need not sacrifice the ability to feed ourselves.

Chris Field, a National Academy of Sciences member, reported in *Science* magazine (2001) that natural ecosystems generally do better than agriculture and other human-managed systems in converting sunlight into living tissue. The plants that anchor those ecosystems have extensive, long-lived root systems with diverse architectures; they have a longer growing season; and their species diversity protects against epidemics and the vagaries of weather. As a result, they can produce, year in and year out, more biomass per acre than agricultural systems without requiring a subsidy of fossil fuels and without degrading soil and water (see color insert *fig*. 5).

The goal of our research team is to develop diverse perennial grain production systems that are as ecologically sound as former prairies. The Land Institute's mission doesn't end at the prairie boundaries of the Appalachians, the Rio Grande, or the Rockies. As Jack Ewel's research in Costa Rica implies, food worldwide can, indeed must, come to be produced by ecosystems that have the efficiency and resiliency of those natural ecosystems that were replaced by farms, forest plantations, and fisheries.

The Land Institute's vision for agriculture extends far beyond the farm gate. Concern is growing that human activity as a whole has become insupportable, the entire planet having fallen into deficit spending, ecologically speaking. But if our species is to find a road leading to great resilience and sustainability, an ecologically sound agriculture can—must—take the lead.

Why should agriculture take the lead? Until now, a feature of agriculture has been to subdue or ignore nature. Yet, ecological processes have long track records of success in building and conserving soil, holding and filtering water, and supporting wildlife diversity.

Agriculture has the discipline of ecology and evolutionary biology to help us produce food in properly functioning ecosystems. All visions of a sustainable or resilient society must rely on renewable resources. Other spheres of human activity do not have that advantage. Agriculture, broadly defined, may be the only artifact in current civilization where that potential resides.

THE ANNUAL REALITY AND THE PERENNIAL OPPORTUNITY

The Midwest contains the best topsoils in the world, yet a growing body of research demonstrates conclusively that the cultivation of annual crops there is degrading soils, rendering water unfit to drink, rolling back biodiversity, spreading toxic chemicals, and even creating a hypoxic, or "dead," zone hundreds of miles downstream in the Gulf of Mexico. The research illustrates a worldwide reality (see color insert *fig. 6*).

Additionally, mountains of evidence show that reestablishing perennial vegetation across the region would solve these problems. But we humans obtain at least two-thirds of our total calories from grains and oilseed crops, none of them perennial. Existing perennial species can produce only a small fraction of the total calories required for direct consumption by a growing human population.

Environmentally conscious researchers and farmers are using the perennial plants available to them, attempting to put more hay and pasture on the landscape. They are planting more trees and grass along rivers and streams to soak up the contaminants that hemorrhage from cropland, and taking more land out of grain production altogether.

In other words, we are treating grain cropping not as a source of life but as a dangerous activity against which humans and nature must be protected.

WHY GRAINS?

European Agriculture Before Coal—Bruegel's *The Harvesters*

The Harvesters by Pieter Bruegel the Elder (see color insert *fig. 7*) was probably painted in July 1565. In this scene of ancient Flemish life, with a large tree in the foreground near which the laborers are eating, drinking, and resting, their entire context seems to be in the wheat field itself. The pear and apple trees are trimmed high, allowing in the light necessary for wheat growth. The wheat is almost as tall as the people, who were probably shorter in that time, but the straw is also longer than most modern wheat varieties. To the right, behind the picnickers, it appears that part of women's work is gathering shocks. Beyond the wheat fields, in the middle background on the left side is a team of horses or mules drawing a wagonload of wheat, presumably to the barn to be threshed. Some of the straw may be used for thatching, but likely most will serve as bedding in the barns both for warmth and to absorb urine and manure. This long-stemmed straw hauled from the barn during winter or early spring will serve as a sponge for returning nitrogen in the form of urine to the fields. David Kline, an Amish farmer friend and writer from Ohio, says this function of the straw makes it about as important to their farms as is the grain.

The standing wheat shocks to the right reveal a relatively small head length compared with the long stem. (A plant breeder of today might say that this crop has a small harvest index, which is a measure of the calorie in the grain.) The stem is cut close to the ground, perhaps because these people of the 1500s wanted to maximize the length of straw for various purposes on the farm or village. Moreover, the crop cuts more easily when the scythe passes close to the ground.

In rural America today, most barns that haven't been torn down need serious repair or will fall on their own. The economic incentive to maintain the barn has mostly vanished. The hay, which used to be housed in the loft, has been replaced by diesel oil or gasoline, which is housed in a metal tank.

Fossil fuel not only led to the loft's obsolescence, it contributed to the reduced need for the lower story, where animals were milked or housed. Commercial fertilizer is manufactured from fossil fuels. With the industrialization of agriculture, energy and nitrogen density greatly increased, and the need to maintain the barn mostly disappeared. Stated otherwise, the painting shows an intact fertility cycle on the farm. Today's farms are on the periphery of our fertility cycle. So, when we see a farm, our eyes might wander to the diesel and anhydrous ammonia tanks and speculate on how long their tenure might be.

PERENNIAL GRAINS RESEARCH AT THE LAND INSTITUTE

When The Land Institute's researchers and their allies around the world are even partially successful in perennializing the major grain crops and domesticating some promising wild perennials for grain, a farm will no longer have to be an ecological sacrifice zone; rather, it will provide food and at the same time protect soils, water, and biodiversity. The missing link—to be persistently effective in using nature as measure—will have been supplied. As those new crops are developed, plant breeders, agroecologists, and farmers will be working out strategies for growing them in mixtures in order to recapture the ecological soundness of preagricultural landscapes.

The genetic raw material is there. Plants now in field plots and on greenhouse benches at The Land Institute and a few allied institutions form much of the foundation of breeding programs that will, given decades of work, turn out perennial grain crops. Most of the current genetic and breeding effort is going into the following species and species hybrids:

Kernza™ is our trademark name for **intermediate wheatgrass** (*Thinopyrum intermedium*). It is a perennial relative of wheat. Using parental strains from the USDA and other sources, we have established genetically diverse populations. We harvested 30 acres in 2009, and an additional 125 acres was planted the same year. There are nutritional qualities superior to those of annual wheat (see color insert *figs. 8–11*).

Wheat has been hybridized with several different perennial species to produce viable, fertile offspring. We have produced thousands of such plants. Many rounds of crossing, testing, and selection will be necessary before perennial wheat varieties are available for use on the farm (see color insert *figs.* 12–13).

Grain sorghum is a drought-hardy feed grain in North America and a staple human food in Asia and Africa, where it provides reliable harvests in places where hunger is always a threat. It can be hybridized with the perennial species *Sorghum halepense*. We have produced large plant populations from hundreds of such hybrids, and have selected perennial strains with seed size and grain yields up to 50 percent of annual grain sorghum's seed size and yield (see color insert *figs.* 14–15).

Illinois bundleflower (*Desmanthus illinoiensis*) is a native prairie legume that fixes atmospheric nitrogen and produces abundant protein-rich seed. It is one of our strongest candidates for domestication as a crop. We have assembled a large collection of seed from a wide geographical area and have a breeding program. We see it as a partial substitute for soybean protein (see color insert *figs.* 16–17).

Sunflower is another annual crop we have hybridized with perennial species in its genus, including *Helianthus maximiliani*, *H. rigidus*, and *H. tuberosus* (commonly known as Jerusalem artichoke). Breeding work has turned out strongly perennial plants. Genetic stabilization is improving their seed production (see color insert *figs.* 18–19).

Upland annual rice, as opposed to paddy rice, is highly vulnerable to erosion. Yet millions of people in Asia depend on it. In the 1990s, the International Rice Research Institute achieved significant progress toward breeding a perennial upland rice using crosses between annual rice, *Oryza sativa*, and the two wild perennial species, *Oryza rufipogon* and *O. longistaminata*. The project was terminated in 2001, but the Yunnan Academy of Agricultural Sciences in southwestern China has been continuing that work with funding support from The Land Institute. The focus is on the more difficult work with the distantly related *O. longistaminata*, which, when crossed with rice, produces plants with underground stems called rhizomes. In recent breakthroughs, a small number of perennial plants with good seed production have been produced (see color insert *figs.* 20–21).

Corn and **soybeans** are two species, one could argue, that more than any other crop need to be perennialized. **Corn** is a top carbohydrate producer. In 2009, eighty million acres were harvested. Until soybean acres increased, corn caused the greatest amount of soil erosion in the United States. It is always number one or two. It will be a challenge to perennialize this crop, but serious consideration is being given to doing so by exploring two main paths: (1) We could obtain genes from a few distant relatives of corn that are in the genus *Tripsicum*. All are perennial, and at least one is winter-hardy. (2) The other, more likely route would be to cross with two much closer perennial relatives of corn. Unfortunately, both species (*Zea perennis* and *Zea diploperennis*) are tropical and not-winter-hardy. Professor Seth Murray at Texas A&M favors using them in crosses rather than *Tripsicum*. Dr. Jim Holland, a USDA corn geneticist at North Carolina State University, says that perennial corn development comes down to a few technical issues that need to be solved. They and other corn researchers are increasing their perennialization efforts.

Several Australian species of the **soybean** genus *Glycine* are perennial; they are difficult to breed with soybean but are potential targets for direct domestication, without crossing to soybean. Our exploration for perennializing soybeans has been very limited. We have been working to make Illinois bundleflower a satisfying substitute.

There is potential for many more perennial grain species, including **rosinseed, Eastern gamagrass, chickpea, millets, flax,** and a range of **native plants.**

TRADE-OFF THEORY

In early November 1982, my sole scientific colleague here at The Land Institute, the late Marty Bender, and I had to answer some scientific thinking for which there was a consensus counter to ours. What follows represents a distillation of some of Marty's notes and a few of mine from that time. Lee DeHaan and Stan Cox, plant breeders here at The Land Institute, more recently have further contributed to the discussion, lending support and a more sophisticated argument.

The idea of an herbaceous perennial grain future understandably prompts the biological question of whether a perennial is capable of a high grain yield. An intelligent, but skeptical, response often wants to compare the morphology of a perennial to that of an annual. The skeptic's intuition concludes that the perennial has to support its roots and/or rhizomes and thus cannot devote as much of its resources to grain as an annual can. This has been expressed theoretically in the concept of r & K selection, where an r-selected species (mice, for example) is considered to be short-lived and devotes a large proportion of its resources to reproduction (such as an annual does), and a K-selected species (elephants, for example) is considered to be long-lived and devotes a small proportion of its resources to reproduction. This trade-off idea began with clutch size in birds. Birds that produced the largest number of eggs had a smaller percentage of their offspring survive to the reproductive years. Those with a few eggs were more protective and had a larger percentage survive to the reproductive years.

The concept of r & K selection matured as a result of the experimental field of resource allocation studies, in which the biomass (which contains energy) content of various organs of a plant or animal collected in the field or grown in a lab is measured as a representation of the allocation of resources within the organism. Thus the typical response to the biological question of perennialism and high grain yield is based on the measurement of production by the organism.

The problem with this response in plants is that it does not consider the processes that limit plant yields; that is, the measurement of production does not say much about the allocation of resources during a plant's growth or about the potential for manipulation of this allocation by plant breeding and agronomic methods. A plant's growth can be limited by energy or by something else, such as minerals. The allocation of energy may not be the most critical limitation to yield, because there is much evidence now that plants normally function at a level of photosynthesis below that of which they are capable. Experimentation has shown that the demand for assimilates by various organs of the plant (sink) has a feedback effect on the rate of photosynthesis (source), with the result that spare photosynthetic capacity may be present, *but not evident unless storage capacity is increased*. For example, grafting experiments with several potato varieties show that tuber size and starch content are largely independent of the characteristics of the aboveground parts. Thus, yield in this case is largely sink-determined.

The seasonal sequence of conditions plays a major role in determining whether source or sink is more limiting. For example, in the cereals, grain filling is largely dependent on photosynthesis and environmental conditions after flowering, but the capacity for storage is determined by the conditions before flowering, so sink can influence source. Plant breeders have demonstrated the reality of this elasticity. If one looks at 1937 data on wheat in Kansas, one can see that the total acreage in Kansas that year was twice what it is today and with half the total yield. Stated otherwise, yields of wheat this year are four times greater per acre than they were seven decades ago.

Here wheat is, civilization's first grain and on the human inventory for ten millennia at least, only now showing its potential, due to breeding over a short period.

Now that the relationship between source and sink has also been revealed, through plant breeding and selection, it is worth remembering that there is no instance where selection for a greater rate of photosynthesis has led to an increase in yield. This illustrates that selection must also be done on the sink. An often-used argument that yield is limited by the source is that there are negative correlations between yield components; that is, if you try to increase one component of yield, another component will decrease. For example, in corn, if more ears per plant are selected for, then fewer kernels per year result. However, an experiment with field beans showed that negative correlations were lowest in the highest-yielding lines and had little to do with establishing actual yield levels. The highest-yielding lines are the best adapted to a limited resource, such as nutrients, that may be causing the negative correlations.

If the relationship between source and sink is limited by resources such as minerals, water, or length of growing season, perhaps this can be resolved by agronomic methods at the field level where perennials have many advantages over annuals:

1. A perennial agroecosystem, like the prairie, has very little runoff or soil erosion, so it has more water and nutrients to utilize than an annual cropping system.

2. Perennial roots have years to grow much deeper into the soil and gain access to nutrients and moisture that annuals cannot reach.

3. Perennial grass roots are the fastest method of building granular soil structure.

4. The initial canopy development of perennials is fast compared with that of annuals, so perennials are able to intercept more light early in the season. They can also continue photosynthesis after harvest, when an annual would be dead.

5. Microhabitats may be present in perennial agroecosystems for organisms such as asymbiotic nitrogen fixers that may not be present in annual agroecosystems.

For these examples one would hope that r & K selection should not be the sole basis for determining whether or not herbaceous perennials can have high grain yields. Here is an analogous example. A warm-season grass species is also called a C4 plant. A cool-season grass species is also called a C3 plant. (The reasons why are at the molecular level and are not worth describing here.) Knowing about these differences in efficiency, farmers and agricultural researchers have not used that sole criterion for determining what crop should be grown in an environment. Furthermore, plant breeding and agronomic methods have been in the array of breeders and farmers for centuries, with enough successes to offset those now willing to feature a sole criterion as to whether perennialism and high yield are actually exclusive.

Humans have forever looked for ways to increase the food supply of the world when all around us, year in and year out, it should be apparent that the most rewarding way would be to increase the proportion of the year when the land is covered by a photosynthesizing leaf canopy.

ECOLOGICAL RESEARCH

Breeding more or less requires that perennials be grown in monocultures (see color insert *fig.* 22). We are also mindful that diversity above and below the surface (see color insert *fig.* 23) has a myriad of processes, most of which are beyond our comprehension. Though breeding is the priority now, we have elected not to wait until perennial grain crops are fully developed to gain experience about the ecological context in which they will grow. At The Land Institute we have established long-term ecological plots of close analogs in which to compare methods of perennial crop management. These perennial grain prototypes, including Kernza and bundleflower, are allowing us to initiate long-term ecological research in these plots. Eventually, improved perennial grain mixtures will succeed them. Additionally, ongoing studies of natural ecosystems, such as tallgrass prairie, provide insight into the functioning of natural plant communities. The prairie is now and likely always will be a valued teacher.

THE ROAD AHEAD

At The Land Institute we have laid out a route to follow in breeding perennial grains and developing the agroecosystems in which they will grow. To expand research on perennial grains across the nation and planet, we freely distribute germplasm—seed of perennials and hybrids. Other plant breeders are using these seeds as parents in establishing or enhancing their own perennial grain programs. Seeds are available for basic research to answer fundamental questions. We are building a body of knowledge about perennial grain systems through publication in the refereed journals.

FREQUENTLY ASKED QUESTIONS

Over the past three decades, interested people have asked some good questions. Here are the most frequently asked, followed by our best answers.

1. It is expected to take at least twenty-five years to achieve more than two or three profitable, productive perennial grain crops. Isn't that too late to address the problems facing the world today?

We hope not, but we do need to move as fast as possible. New strategies are needed that emphasize efficient nutrient use in order to lower production costs and minimize negative environmental impacts. The sooner that successful alternatives are available, the more land we can save from degradation. It is likely that global agricultural acreage will expand over the next two to three decades, especially if the human population increases to eight to ten billion people. Recent projections predict an 18 percent or more increase in agricultural land by 2020. The best soils on the best landscapes are already being used for agriculture. Much of the future expansion of agriculture will be onto marginal lands (Class IV, V, and VI) where risk of irreversible degradation under annual grain production is high. As these areas become degraded, expensive chemical, energy, and equipment inputs will become less effective and much less affordable.

Thirty-eight percent of global agricultural lands are currently designated as degraded, and the area is increasing. To minimize encroachment onto nonagricultural lands in the future, currently degraded lands will need to be kept in production *and* restored to higher productive potential. In regions of the world where high inputs of fertilizers, chemicals, and fuels are not an option, agricultural systems that are highly efficient, productive, and conservative of natural resources are needed—and will be needed even more twenty-five years from now.

2. Can we expect perennial grain crops to be as productive as annual grain crops, and if not, won't they actually worsen environmental problems by requiring more land for agricultural production?

a. There is sufficient evidence that "reasonable reference yields" of annual crops can be matched on high-quality lands and exceeded on poor-quality lands by diverse perennial systems with fewer negative impacts.

b. It depends on which annual yields are used as a standard. For example, the world record wheat yield was harvested in the Palouse region of Eastern Washington State, where wheat yields can top a hundred bushels per acre. Annual wheat production in that region, though, has resulted in extensive erosion. All of the topsoil has been lost from over 10 percent of the region's landscapes. On eroded sites Palouse wheat yields may be less than twenty-five to thirty bushels per acre. Crop yields that come at such a high cost to the soil resource—or that depend on an extravagant use of chemical fertilizers—should not be used as a standard of comparison.

3. But won't the seed yield of perennials always be limited by the need to save some energy for overwintering that could have been used to produce seed?

The short answer is no. The theoretical limitations to seed yield in perennials are no more serious than in annuals. In annuals, yield is limited by shorter growing seasons, water shortage due to short roots, and poor seedling establishment. In perennials, first-year-of-growth yield can be constrained by the need to overwinter or oversummer, but rapid growth of perennials in their second and succeeding years of growth, combined with season-long access to water deep in the soil profile, means that perennials such as alfalfa are overall more productive than related annuals like soybeans. Much of the work of plant breeders has been to shift the allocation of resources from leaves, stems, crowns, and roots toward seed in the development of perennial grain crops.

4. With advances in no-till production of annual grain crops, do we need perennial grains to mitigate the environmental problems associated with agriculture?

Unfortunately, yes. Although no-till technology has reduced erosion in many areas, some problems remain due to the biological limitations of annual plants. Chief among the problems associated with no-till is water quality. Annual crops, even in no-till situations, are relatively inefficient in capturing nutrients and water. In the Midwest, as much as 45 percent of precipitation may be lost through the soil profile under annual cropping. Rates of water loss through profiles may be five times greater under annuals than under perennials. No-till, compared with conventional tillage systems, can have losses as great or greater.

Annual crop plants are either absent or too small to use and manage water during times of precipitation. Water flowing through the soil profile transports soil nutrients and agrichemicals downward. Poor water quality is the consequence. This problem can be compounded under no-till production, which often requires greater inputs of agrichemicals and fertilizers. A 2002 EPA survey of the nation's water quality shows a downward trend from the late 1990s. The problem is getting worse, despite widespread adoption of no-till and minimum-till systems.

Crop seeds need warm, well-drained seedbeds in order to properly germinate. No-till limits this. That is why tillage remains an attractive practice in northern regions. Warming and drying of the seedbed can be hastened. Advances in plant breeding may eventually allow for optimal germination in cooler, wetter conditions, but in the Midwest, seedlings will still be small when the rains come.

5. If our farming systems "mimic," to some degree, natural ecosystems, what level and kind of plant diversity are needed and how will it be deployed?

The answer to both parts of the question is "It depends." It depends on the resilience and fertility of the soil, climate, disease pressures, and types of crops. Nearly all of nature's land-based ecosystems feature perennial plants grown in diverse mixtures. Natural ecosystems, in general, use and manage water and nutrients most efficiently, and build and maintain soils. For that reason alone nature is our standard. The level and spread of diversity vary. The characteristics of the region in which the plants are to be grown will have to be assessed.

Diversity is of two kinds: multiple species and genetic diversity within species. Current grain production practices commonly involve planting a single genotype (near-zero genetic diversity) across a field often larger than a hundred acres. Furthermore, that single genotype and other genetically similar plants are being grown on millions of acres in a region. Increases in genetic diversity at the species, field, and landscape levels are needed. The final ordering of the components of that diversity will be determined by what is useful and can be practically achieved by local farmers.

6. Several serious attempts have been made in the past to perennialize grain crops, and we have none to date. What has changed that offers promise of success now?

History need not be a source of discouragement. In the case of wheat, most involvement with perennials had to do with bringing desirable genes—for resistance, say—from a wild perennial relative into the annual crop. The perennialization effort in most cases was carried on more or less as a hobby, by an interested researcher but with no institutional commitment for a sustained program to guarantee continuity. When the researcher retired, the effort ended. The Soviets had the most ambitious perennial wheat program, but political decisions halted these efforts in the late 1950s or early '60s.

We are now in a new era in two ways:

a. In recent years, the costs to our soils and waters due to annual cropping are increasingly weighed against bushels per acre, making some reduction in yields acceptable.

b. With recent advances in plant breeding, more knowledge of the genome, and greatly increased computational power, thinking about breeding limits has changed.

7. Since mechanical tillage and annual rotations are largely eliminated in perennial systems, don't the perennial plants become "sitting ducks" for pests and disease?

Here proof is in the pudding. Perennials dominate most native landscapes and constitute roughly 80 percent of North America's native flora. Perennials have thrived throughout the evolutionary history despite the pressures of pests and disease.

In some fields or some regions, some perennial crops will prove to be more problematic than others, and breeding for complex traits like yield and perenniality can unintentionally purge genes involved in resistance responses. There will undoubtedly be pest and disease problems. But these problems also afflict our most productive annual crops. And there are many examples of herbaceous perennial plants—alfalfa, switchgrass, brome—that remain highly productive for many years despite exposure to pests or disease. Diversity (whether at the field or landscape scale or over time), field burning, and selecting for resistance in a plant breeding program are essential elements of our work.

8. How do alternative methods of production such as permaculture, bio-intensive, or organic fit in with perennial grain crops? What about vegetables and fruits? How do community-supported agriculture farms fit in?

We focus on the crops that occupy 68 percent of global cropland and provide about the same percentage of food calories: annual grain crops grown primarily in monocultures. Any number of approaches, alternative or conventional, could be used in managing perennial crops and distributing the harvest.

This is not to say that efforts aimed at reducing the scale of industrial agriculture and increasing local food security are misguided. They are not! They are necessary to transform our food system over the long term. While promoting local, small-scale, organic agriculture, we must also assess how and where the bulk of our calories can best be produced. If all or even a large portion of the calories consumed by New Yorkers came from New York State, there would be few trees left, and the state's thin, poor soils would be quickly degraded. The bulk of the calories consumed by New

Yorkers must come directly or indirectly from grain crops that grow well in the Midwest and Great Plains states.

9. Will the public eat perennial grains?

People have liked the rolls, breads, pancakes, etc., that we have made with Kernza (a perennial close relative of wheat), and we see little reason for people to find significant or undesirable taste differences in perennial grains generally. Greatest short-term success in developing suitable perennial crops will come with perennializing current grain crops with which the public is already familiar. Indeed, one of the strongest arguments for perennializing those grains is that it does not require large dietary shifts.

10. How are you going to harvest a perennial grain polyculture?

This question arose so frequently over the years that we finally decided to plant a polyculture of four annual crop species: corn, soybean, sorghum, and sunflower. The seed mixture was planted with an air drill. At harvest we opened the concave on the combine and cut the air (so as not to blow the sunflower seeds out the back). Progress through the field was slow, but not prohibitively so. Seeds were separated with a seed cleaner. The point is that mechanical equipment already in existence, with a little fine-tuning, can do the job. The larger problems are agronomic, not engineering.

11. If perennial grain crops are highly productive and become widely planted, won't they require even fewer people on the landscape and thereby worsen rural community life?

Maintaining high seed yields in these systems will require greater skills based on an understanding of the climate, soils, and productive capacity of a particular place. Some type of rotational management may be required where, for example, in one year cool-season grasses are simply grazed by livestock and the warm-season plants are harvested later in the season for their seeds. The following year, two seed harvests might be possible. The year after that, perhaps the entire field is grazed the whole season. A whole array of weed, disease, and fertility challenges will require well-informed, skilled management. Knowing when to burn, graze, harvest, or let lie fallow, and knowing when and how to monitor for pests, disease, and weeds, will require *more* attention than current agricultural production (see color insert *figs.* 24–27). A sufficiency of people will substitute for a sufficiency of capital in a world of resource scarcity. But it need not be thistles and thorns and the usual "sweat of the brow" described by the Genesis mythmakers. The psychology will be more like that of a nineteenth-century British naturalist than of the modern-day grain-producing farmer. Roughly 90 percent of current annual farm revenue goes off-farm, requiring increasingly larger farms to generate enough income for a family. With a greater percentage of financial returns going to the farmer, fewer acres will be necessary to comfortably support a family and still send food to nonfarm families.

WHAT ABOUT BIOFUELS FROM
PERENNIAL POLYCULTURES?

Now and then, interest is shown in perennial polycultures as a source of biofuels for our internal combustion engines. Jerry Glover, soil ecologist at The Land Institute, has responded to this interest. He began by acknowledging that the science is not yet there for the cellulosic ethanol (EtOH) process. But assume the current cost estimate is five to six times that of corn EtOH plants. Jerry assumed "the day will come when it will be." What will be the role of perennial polycultures? First off, will an industrial process with very high capital costs and technological sophistication and *with a need for a steady supply of feedstock* be compatible with the current U.S. farming economy? If so, what will be the energy cost of adjustment?

These are some positive considerations:

1. Many grasslands developed under a regime of biomass removal. We have grazers and hay meadows with complex species mixtures that could work as biofuels (see color insert *figs.* 27–30).

2. Biofuels may make control of exotic problem species such as johnsongrass or *Sericea lespedeza* cost-effective.

3. Cellulosis biofuels are potentially compatible with livestock, honey, wildlife (game and nongame), conservation, *and* wind farms.

4. Perennials in monocultures have the advantage over annuals of sustained yield on marginal landscapes. We make up for this deficiency with nonrenewable inputs.

Most of the above virtues are due to more rooting volume and a longer growing season.

Perennial *mixtures* have even better resource access and management. Annual crops typically recover less than 50 percent of fertilizer nitrogen, making more fertilizer necessary to offset the season's losses. Nitrogen fertilizer is very energy-intensive. For years it was two times more than the energy spent for traction on U.S. farms.

Some Possibilities

The land devoted to the Conservation Reserve Program is now being considered as a candidate for biofuels. That program cost our federal government an average of $500 per acre over a ten-year period. This thirty-six million acres, roughly one-tenth of our agricultural land (not counting rangeland), needs to remain in cover to save the soil resource. Even corn stover and other crop residues should not be regarded as a "waste looking for a place to go." It is important to remember that carbon left in the field for soil health will be more important in the field than in the gas tank.

In the past thirty years, our long-term research in perennial mixtures has taught us much. These perennial mixes have promise as a carbon sink (though once only) and as the most benign renewable energy source from the land.

Native prairie sequesters more carbon than cropland, and using prairie plants for energy rather than gasoline or diesel could result in six to sixteen times fewer greenhouse gas emissions than using corn ethanol to replace gasoline or diesel. This is because prairie plants absorb more carbon dioxide from the atmosphere than was produced to plant, fertilize, harvest, and process them for energy. Finally, as David Tilman notes, prairie plants could be 51 percent higher than net returns from corn grain ethanol.

Because The Land Institute is in Kansas, Jerry Glover went on to address what this means for Kansas as follows:

1. Kansas has around 54 million acres. A bit more than 2.6 million acres is currently enrolled in the Conservation Reserve Program (CRP). Much wildlife habitat has been restored, and there has been reduced soil erosion.

2. Even greater conservation achievements could be attained by also converting highly erodible lands (HEL) currently used for crop production to perennial native plants for bioenergy. HEL

acreage is at a greater risk of being eroded and degraded when used for annual crop production such as corn. Such acreage typically produces lower crop yields or requires more chemical inputs (more fossil fuel input as well as pesticide and nitrogen toxicity) to maintain yields.

3. Combined, CRP and HEL acreage represents *20 percent* of Kansas land area, about 10.7 million acres. If the 10.7 million CRP and HEL acres were used for bioenergy production, they could potentially be used to meet 14 percent of the state's current total energy use. That is being very optimistic because

   a. this method assumes that an advanced, highly efficient method (using integrated gasification and combined cycle technology with Fischer-Tropsch hydrocarbon synthesis) will be used to produce gasoline and diesel as well as electricity.

   b. this estimate optimistically assumes prairie hay yields around two and a half tons per acre.

   c. we need 20 percent of Kansas land.

Two lessons to be drawn from these three factoids:

• The rate of increase in liquid fuel use can wipe out any significant gain.

• Turning the agricultural landscapes of the world into fuel for automobiles, trucks, and airplanes as a massive fuel program can never do what an extremely modest conservation program can do.

LANDSCAPE ECOLOGY

Small Farms and Social Justice: A Necessity
to Save the Wild Biodiversity in the Tropics

"Agriculture is the number one threat to biodiversity," says the Millennium Ecosystem Assessment report.

Nature's Matrix, by Ivette Perfecto, John Vandermeer, and Angus Wright, carries the argument that biodiversity conservation in the tropics— home to most of the world's species—cannot succeed by creating protected areas alone. Ecological thinking, some old, some new, now holds:

1. Local extinctions can be expected in most species, whether humans are there or not.

2. The ability of the larger surrounding landscape to accommodate migration and reproduction in new locales is essential since the setting aside of wild corridors to connect wild landscapes has proven unable to do this.

3. In this tropical area a large share of the landscape is in agriculture. Therefore, the key to biodiversity conservation will depend on agricultural landscapes diverse enough to support migration and propagation of species *outside* the protected areas. The authors call this a "high quality matrix" and necessary for species survival.

4. Diverse agriculture is mostly in the tropics, more or less totally managed by traditional and/or small-scale farmers.

5. These farmers are protected and promoted by various social movements that have recently adopted the phrase *food sovereignty*, which covers descriptions of their key goals. So, in summary, we are forced to conclude that, yes, biodiversity conservation will require clearly identified protected areas. But it will require also an alliance of conservationists with people and

social movements that create or maintain diverse, mostly small-scale agriculture.

Here we see that the ecosystem concept has the power to bring conservation, agriculture, and social movements together. This is a wonderful thought, and one cannot help but wonder what the implications might be for California, Kansas, and Vermont. Here the problem becomes more complicated because landscapes in the developed world have less biodiversity. Does that mean that social justice in the Central Valley of California is irrelevant since the wild biodiversity is gone? Is there enough diversity in the refugia to invite them back? Or can a diversity of domestic species invite them back?

Chapter Eleven

A 50-YEAR FARM BILL PROPOSAL GOES TO WASHINGTON

Wendell Berry, Fred Kirschenmann (Leopold Center for Sustainable Agriculture, Iowa State University), and I took what follows in this chapter to Washington, DC, and delivered it to Deputy Secretary of Agriculture Kathleen Merrigan in July 2009.

INTRODUCTION

Long-term food security is our issue. We begin with the knowledge that essentially all of nature's ecosystems feature perennial plants growing in species mixtures, and that they build soil. Agriculture reversed that process nearly everywhere by substituting annual monocultures. As a result, ecosystem services—including soil fertility—have been degraded. Most land available for new production is of marginal quality that declines quickly. The resulting biodiversity loss gets deserved attention, soil erosion less.

ACKNOWLEDGMENT OF COALITIONS

To address diminishing agricultural potential with a new vision, The Land Institute sponsored ten meetings coast to coast with farmers and representatives of sustainable agriculture organizations. This loose coalition can help to build a broader constituency.

Organic and local food organizations, including some represented in our coalition, provide vision, education, and models of greater sustainability. With those constituencies, we share common principles and the goal of returning the world's grain-producing landscapes to perennial plants in the rotation for grain production.

Green Lands Blue Waters is an Upper Midwest coalition advocating the need to perennialize the landscape of the Mississippi Basin out of concern for soil erosion and the leaching of nitrogen, which is responsible for one of the largest dead zones of the world. GLBW partners include the University of Illinois, Iowa State University, the Leopold Center for Sustainable Agriculture, Louisiana State University, the University of Minnesota, North Dakota State University, the University of Wisconsin, the Audubon Society, the Illinois Stewardship Alliance, the Institute for Agriculture and Trade Policy, The Land Institute, The Land Stewardship Project, the Minnesota/Iowa Farmers Union, The Nature Conservancy, Trout Unlimited, Practical Farmers of Iowa, and the Rural Advantage and Agricultural Watershed Institute.

WHAT IS REQUIRED?
PROMOTE SYSTEMIC CHANGE

A 50-Year Farm Bill is a proposal for gradual systemic change in agriculture. Perhaps what has been missing is an available vision with scientific feasibility. Implementation will depend on endorsement by the secretary of agriculture, the president, Congress, nonprofit organizations, corporations, and citizens.

PLAN

Enclosed are charts that illustrate changes over ten five-year farm bill periods. Each five-year bill, in addition to its existing programs for subsidies, food programs, etc., moves incrementally toward the fifty-year goal of stopping the deficit spending of ecological capital necessary for food production. Thus, the 50-Year Farm Bill becomes an instrument for increasing sustainability and food security.

In the short run, we can achieve a significant measure of success through farm policy that encourages farmers to increase the use of perennial grasses and legumes in crop rotations. But that will not be enough. Options for farmers will take a major leap when perennial grains are available. Their input costs will decline as the landscape benefits. USDA and other researchers will need policy to sustain funding. Breeding perenniality into a broad spectrum of our current grain crops will take time. Even so, prototypes have thrived for several years in Kansas. As their yields increase, they will replace their annual relatives—one in as few as ten years.

Our project would employ the ecosystem as the standard. Once that standard is adopted, an array of technologies can become useful tools. Technology would follow, rather than lead, the vision.

COST
USDA FUNDING

We do not seek USDA funding for The Land Institute or any particular organization. The Land Institute will offer to the project free germplasm and more than thirty years of experience with perennials. Its staff in this decade has greatly enhanced the diversity of crops and speed of change. We have hybrid prototypes of perennial wheat, sorghum, sunflower, and other crops. We are giving people small samples of flour from a perennial wheat relative we have named Kernza™. Biochemical analysis shows it to be superior to annual wheat in nutrition. People like it. We expect it to be farmer-ready in a decade.

During three decades, we have collaborated with several land-grant universities and other institutions. We include them as assets. **Because the change needed is systemic, we believe that USDA should take the lead.** The Obama administration's devotion to change makes our proposal now seem possible.

We propose that, over an eight-year period, federal funding would sponsor eighty plant breeders and geneticists who will develop perennial grain, legume, and oilseed crops, and thirty agricultural and ecological scientists who will develop the necessary agronomic systems. They will work on six or eight major crop species at diverse locations. Budgeting $400,000 per scientist-year for salaries and research costs would add less than $50 million annually. This is 8 percent of the amount that the public and private sectors have been spending on plant-breeding research alone, according to a late 1990s survey.

REVERSING ECOLOGICAL DAMAGE

Our vision is predicated on the need to end the ecological damage to agricultural land associated with grain production—damage such as soil erosion, poisoning by pesticides, and biodiversity loss. The most cost-effective way to do so and stay fed is to perennialize the landscape.

The transition of agriculture from an extractive to a renewable economy in the foreseeable future can now be realistically imagined. Our proposal is ambitious, but it is necessary, and it is possible. We have little doubt that we can make the agricultural transition faster than the adjustments imposed upon us by climate change and the end of the fossil fuel era. If we can keep ourselves fed, we have a chance to solve the other problems.

CONCLUSIONS

Perennialization of the 70 percent of cropland now growing grains has potential to extend the productive life of our soils from the current tens or hundreds of years to thousands or tens of thousands. New perennial crops, like their wild relatives, seem certain to be more resilient to climate change. Without a doubt, they will increase sequestration of carbon. They will reduce the land runoff that is creating coastal dead zones and affecting fisheries, as well as save and maintain the quality of scarce surface and ground water. U.S. food security will improve.

Social stability and ecological sustainability resulting from secure food supplies will buy time as we are forced to confront the intersecting issues of climate, population, water, and biodiversity.

Five-year farm bills address exports, commodities, subsidies, some soil conservation measures, and food programs.

A 50-Year Farm Bill would be a program using five-year farm bills as mileposts, adding larger, more sustainable end goals to existing programs:

- Protect soil from erosion.
- Cut fossil fuel dependence to zero.
- Sequester carbon.
- Reduce toxins in soil and water.
- Manage nitrogen carefully.
- Reduce dead zones.
- Cut wasteful water use.
- Preserve or rebuild farm communities.

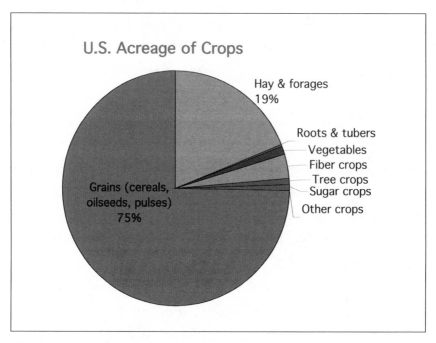

Figure 17.

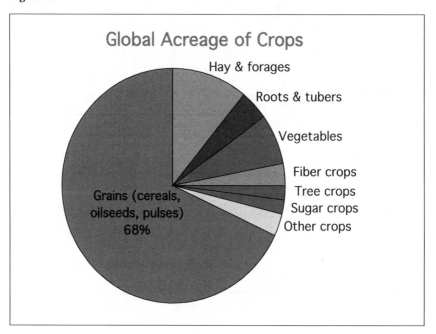

Figure 18.

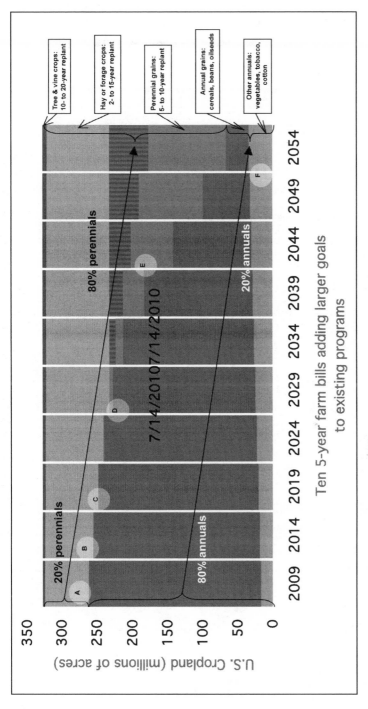

Figure 19.

A, 2009: Hay or grazing operations will continue as they exist. Preparations for subsidy changes begin.

B, 2014: Subsidies become incentive to substitute perennial grass in rotations for feed grain in meat, egg, and milk production.

C, 2019: The first perennial grain, Kernza™ (a wheat), will be farmer-ready for limited acreage.

D, 2029: Educate farmers and consumers about new perennial grain crops.

E, 2044: New perennial grain varieties will be ready for expanded geographical range. Also potential for grazing and hay.

F, 2054: High-value annual crops are mainly grown on the least erodible fields.

U.S. AND GLOBAL CROPS

Although we start with our own country's soils and food supply, negative results of our present agriculture—soil erosion, chemical contamination, fossil fuel dependency for food production, and dead zones—are global problems, so this 50-Year Farm Bill ultimately is for the world.

PROTECTING OUR SOIL WITH PERENNIALS:
NATIONAL ACREAGE GOALS

Half a century of concerted investment in research, education, and incentives to conserve soil with deep-rooted, long-lived perennial crops could increase the protected acreage from 20 to 80 percent.

Pastures and perennial forage crops are already available either in permanent stands or in rotations. We propose incentives that would maintain the present perennial acreage and increase perennials in rotation. When perennial grains become available, they will require no financial subsidy, since they would represent a compelling alternative.

Figure 19 projects what is possible.

Components of agricultural sustainability

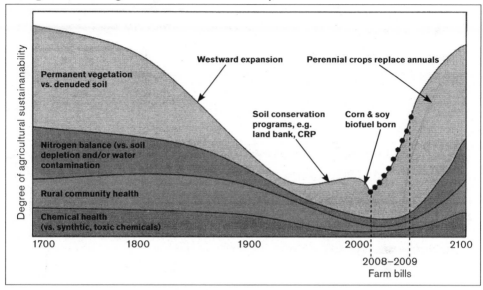

Figure 20.

Changes in USDA program priorities to increase the productive lifespan of US cropland

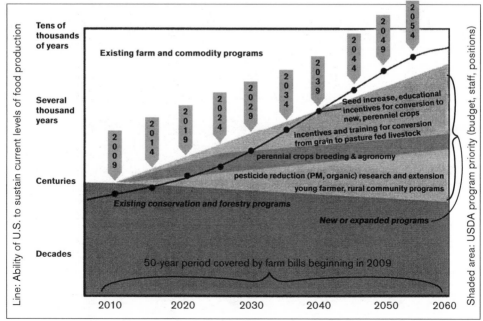

Figure 21.

A 50-YEAR FARM BILL GOES TO WASHINGTON

Prudence requires one to accept the Millennium Ecosystems Assessment conclusion that agriculture is the "largest threat to biodiversity and ecosystem function of any single human activity." In the United States it is responsible for 70 percent of water contamination. Pesticides are in nearly every water and fish sample from agricultural areas. Some 40 percent of U.S. waters are unfit for swimming and fishing.

With global agricultural expansion is expected an 18 percent increase in cropland, a 30 percent increase in fertilizer, a 75 percent increase in pesticide production, a 250 percent increase in eutrophication, and rising greenhouse gas emissions. Most of the expansion will take place on soils less resilient, and there will be further soil degradation. There will also be further loss of biodiversity, reduced pollination, and loss of critical ecosystem services, including serious compromise of water and nutrient cycling. Natural biocontrol will decline.

Harold Mooney, et al. (2005), concluded that saving wild biodiversity boiled down to "intensive management with high yields, versus . . . lower yielding systems." In other words, *production at the expense of conservation* or *conservation at the expense of production*. The dichotomy is correct for now because all grains are annuals.

It will take time, but the safe path is to adopt the ecosystem concept precisely because ecosystems have been successful micro-managers of landscape for millions of years. Nearly all feature perennial roots of varying architectures, alive year-round, binding the soil, making it less susceptible to wind and water erosion. Their greater access to water and nutrients over a longer growing season grants them excess capacity that a plant breeder can reallocate to grain production.

We have the ability to cross certain domestic annuals with wild relatives and domesticate some promising wild perennials. The transformation of wild species into crops has been done before—with annuals. *Conservation as a consequence of production* answers the dilemma laid out by the Millennium Ecosystem Assessment.

A final contrast tells the story. Illinois has been called The Prairie State. It once was dominated by warm season tall-grass species. Now we have supplanted huge acreages of that prairie with soybeans and another warm season tall grass, corn. When we see that flat Illinois corn field eroding (see color insert *fig.* 31), it is not tall grass or warm season we should blame, but the annual habit. The tall-grass species still intact over hundreds of thousands of acres like these in the Kansas Flint Hills (see color insert *fig.* 32) feature a diversity of perennial roots. The risk is to not act on what is apparent to the casual observer. What is missing is the political will.

CHAPTER TWELVE

AN APPEAL TO
THE RUSSIANS

At the end of August 2009 in Coon Rapids, Iowa, there was a celebration of the fiftieth anniversary of Nikita Khrushchev's visit to the Roswell Garst hybrid seed farm. Sergei Khrushchev, now a U.S. citizen, was there as he had been fifty years earlier with his father. There were twenty-five or so Russians present, including the secretary of agriculture and the ambassador, plus some two hundred Americans, including representatives of U.S. agribusinesses. I was asked to give a talk. I thought this a good opportunity to invite the Russians to a cooperative effort on the substance of the 50-Year Farm Bill. In preparation, I had read some of Khrushchev's memoirs, noting especially the time of the visit with the Garst family, plus much that has happened since. What follows is that talk.

The Next Fifty Years: U.S.-Russian Cooperation on Perennializing the Major Crops with the Ecosystem as the Conceptual Tool

September 23, 1959. A pivotal moment in history? For us assembled here, it is. We have gathered today to think about that time fifty years ago and all that has happened since. It is also worth thinking backward a hundred years, from 1959 to 1859, when Colonel Drake's oil well was drilled in western Pennsylvania. That was the first one ever. 1959 was also the hundredth year since Charles Darwin's *On the Origin of Species* was released. And 1959 was the fiftieth anniversary of the most important invention of the twentieth century, the Haber-Bosch process, which made it possible to turn atmospheric nitrogen into ammonia fertilizer. According to energy scholar Vaclav Smil, without this process 40 percent of humanity would not be here today.

Neither Nikita Khrushchev nor Roswell Garst nor most anyone else could imagine the speed of change to come over the next fifty years. They could not know, for example, that a ten-year-old in 2009 would have lived through the burning of 25 percent of all the oil ever burned. How could Khrushchev or Garst or anyone else have imagined the scale of the consequences of burning the manure of the farmstead and substituting chemical fertilizer, as Mr. Garst advocated? Mr. Khrushchev thought the idea a good one. None of us, I'm betting, could have foreseen the number of nitrogen-polluted wells in rural America or the cost, for cities such as Des Moines, of removing nitrogen from the public water supply. Nor could we have foreseen that nitrogen runoff would cause dead zones in the oceans.

The industrial revolution was a bit over two centuries old when the early stages of agriculture's Green Revolution were getting under way in 1959.

189

Within three decades, yields of several major crops had doubled in some places, tripled in others. Oil supplied the fuel for traction and other mechanical operations. Natural gas sponsored the ammonia to fertilize the crops that realized the yield potential developed by the plant breeders. Every gain in bushels per acre or kilograms per hectare was praised. As the food supply went up, antibiotics and other drugs forced death rates down. The world population doubled. By means of all these successes, what we can call the Industrial Mind now permeates the world.

As fossil fuel consumption accelerated, only a handful of scientists speculated on the costs to the ecosphere of the released carbon. But now global warming is generally recognized as a public enemy.

Nikita Khrushchev and Roswell Garst did know one big thing, as true today as it was then: people will want to be fed, every day, every week, every month, every year. This need must have been most poignant for Mr. Khrushchev. The Siege of Leningrad and the resultant starvation, sickness, and death lasted from August 1941 until January 1944, only fifteen years before the Iowa meeting. No matter how long and dark the tunnel might be for countless problems sure to face us, both the farmer and the premier knew that if we could produce enough food, most problems would be manageable. Their question then is our question today: How do we help ensure an adequate food supply, not just in our two countries, but around the world and for centuries to come? Next year our earth is expected to feed seven billion people.

Solutions to the energy and climate crisis will require extraordinary political will, both to conserve energy and to develop renewable energy technology. We might succeed. But once soil has eroded, there will be no technological substitute. And in spite of all our efforts so far, soil erosion and other landscape degradations are increasing globally. In a few places they have been slowed by minimal-till or no-till farming, but with this so-called fix, pesticides accumulate. We are poisoning our soils to save them.

I want to propose a solution now that both of our countries can embrace. If we cooperate, success will come faster. Annuals grown in mono-cultures dominate agriculture. But essentially all of nature's ecosystems feature perennials growing in mixtures. We need to breed perennial grains. The literature of ecology and evolutionary biology has been accumulating on the shelf for a century, more or less for its own sake. It is a great untapped potential to draw on. Ecologists have learned to detect many of the efficiencies of the various ecosystems, both above and in the soil. The ecosystem becomes our conceptual tool. Nature's way of operating on the land becomes the standard against which we judge our agriculture.

Imagine two ends of a spectrum—human cleverness at one end and nature's wisdom at the other. The Industrial Mind, over the last hundred years, has increasingly relied on human cleverness. I am not proposing that we quit being clever, but instead, cleverness should be subordinated to nature.

Advancing this idea will be a great challenge. Our ten-thousand-year history of growing food has been tied to the notion that nature is to be subdued or ignored. But this attitude has made for relentless deficit spending of the earth's capital. Now, finally we face the need to change course. The possibility of doing so resides in exploring the efficient processes of nature, which are sponsored by contemporary sunlight.

This look to nature begins with the soil itself. For tens of millions of years, nature's arrangements have managed the twenty-some elements that go into all organisms. Only four of these elements—carbon, hydrogen, oxygen, and nitrogen—are found in the atmospheric commons. Our future lies in the soil.

This brings me to why Russians need to be major players in this effort for a new agriculture. I have five stories out of Russia's rich culture. Three are about scientists, one is about a moment in history, and the fifth is about a historical period. After I have told all five stories, I will explain what our two countries could accomplish over the next half century, to 2059.

1. Vasily Dokuchaev is the father of soil science. In the 1870s, Dokuchaev was given the task of describing the structure, origin, and evolution of the deep, rich grassland soils of western Russia. Classifying them had been elusive. The belief of the time was that weathering alone was responsible for soil formation. Soils were thought to have no emergent properties of their own, no properties due to interactions with organisms, nothing that would give them standing beyond what mere weathering would cause.

To accomplish this task of soil classification, Dokuchaev traveled over ten thousand kilometers (six thousand miles). From his observations he concluded that "soil exists as an independent body and has its own special origin and properties unique to it alone." More is involved in soil formation than merely moisture and temperature. He identified five factors that govern formation of soil: climate, parent material, organisms, topography, and time. His conclusions revolutionized soil science, earning him the title of father of soil science. Lacking translation from Russian, Dokuchaev's idea was not recognized in the West for decades. But when Professor Hans Jenny, father of American soil science, published his seminal book, *Factors of Soil Formation*, in 1941, he accepted and used Dokuchaev's five factors. Professor Jenny's contribution was to create a formula that required differential equations. By providing quantitative measures of the five factors of Dokuchaev, Jenny began a revolution in soil science.

2. Dr. Nikolai Ivanovich Vavilov, born in 1887, was internationally known as an agronomist, botanist, plant breeder, geneticist, and plant geographer. He had a big-picture view of our earth, both geographically and in time. Early in his career he set out to determine where the cultivated plants originated, traveling worldwide and making massive collections. His published conclusions set the standard for all subsequent investigations. Plant breeders still turn to them again and again. His passion and intellect were supported by a rich Russian culture notable for its love and honoring of natural history.

3. Theodosius Dobzhansky was a Ukrainian immigrant who came to the United States, became a citizen, and is regarded as the most important evolutionary biologist of the twentieth century. In 1937, twenty-two years before Khrushchev's 1959 visit, Dobzhansky published his landmark book, *Genetics and the Origin of Species*, in which he bridged a wide gap between

Figure 22. N. I. Vavilov on the Papago Reservation (photo by Homer Schantz)

experimental geneticists and naturalists. He was the first to successfully integrate understanding of evolutionary problems, from his naturalist perspective gained in the Ukraine, with experimental genetics, from his more recent experiences in the United States. The great American evolutionary biologist, the late Ernst Mayr, said that Dobzhansky's 1937 synthesis, so long in arriving, was the most decisive event in biology since 1859, the year of Darwin's *On the Origin of Species*.

Now for two important times in the history of Russian culture.

4. During the 872-day Siege of Leningrad, not tens but hundreds of thousands of people died from hunger. There was every reason to believe— given its huge quantity of seeds collected from all over the world—that the Institute of Plant Industry would be overrun by the hungry people. But the institute's staff guarded the collection. And in the midst of the seeds, they too starved. The collection remained untouched.

5. Finally, there is the extensive work carried out in Russia on the remote hybridization of plants. A team of plant breeders made countless wide crosses between species and varieties. Their desire was to speed crop plant evolution. According to N. V. Tsitsin, one of their efforts was to develop a wheat with "perennial character, remarkable for dwarfness, resistance to lodging, an ear structurally similar to that of conventional wheats, and easy threshability." He also said that "among the primary features of the perennial and feedcorn wheats we are currently putting through selection is *their ability to develop a powerful rooting system*, a factor rather important for the maintenance and betterment of the soil structure, but one of prime importance for the regions susceptible to wind erosion" (emphasis mine). Tsitsin acknowledged that "it

will be some time before the newly bred varieties can be cleared for production, lagging far behind the best wheat varieties in terms of their yields."

He went on to say, "The fields under perennial wheat need not be reploughed, the stubble will be highly snow-retentive and thus very likely to promote the accumulation of soil moisture. In sum, these factors will provide for *sustained and progressively increasing fertility of the fields*" (emphasis mine).

Of course, one wonders what happened. Why did the work on perennial wheat not continue? Perennial wheat development was part of a larger program in the Soviet Union. It was pursued, more or less, as a sideline, because: (1) farmers and governments need ensured high yields every year, so the bulk of research funding went to that end, and (2) any long-term research effort in agriculture is the first to be cut when funding is reduced.

Breeding perennial grains to gain high yield is a long-term venture. Abundant fossil fuel inputs in agriculture were then available to offset some of the consequences of soil erosion. Fossil fuels make it possible to mine, package, and transport nutrients from afar. Commercial nitrogen is made possible by the energy-intensive Haber-Bosch process. For whatever reason, the perennial wheat breeding faded away before it was farmer-ready.

In Conclusion

Those five stories bring us to the modern challenges of achieving sustainable food production. We need to

• feed the expanding population as we work to stop its growth.

• prevent soil erosion beyond natural replacement levels.

• manage nutrients and water more efficiently.

• greatly reduce the use of toxic chemicals.

A Proposal for a Cooperative Effort

I propose that our two countries jointly undertake a massive breeding program devoted to perennializing the major crops currently responsible for occupying at least two-thirds of the agricultural land of the planet and responsible for two-thirds to three-fourths of our calories: mostly grains and pulses (legumes). Joint ecological studies of agricultural landscapes could get under way at the same time.

The vision and the work of several organizations interested in a resilient agriculture have now produced a 50-Year Farm Bill, which proposes a gradual, systemic change in U.S. agriculture. It is a scientifically feasible plan for preserving the landscape for food production. This plan, at a minimum, will shape the future work of our coalition. We offered it to our secretary of agriculture, Tom Vilsack, with the hope that he and his department will see the need for it.

Until the plant breeders develop farmer-ready perennial grains, ecologists can use analog species to experiment with various mixtures, various root architectures, yielding various ecological functions. At the outset the proposal advocates more perennials in rotation, and more pasture. It further states that our five-year farm bills should serve as mileposts of progress over fifty years to move the U.S. agricultural land from 80 percent in annuals and 20 percent in perennials to 20 percent in annuals and 80 percent in perennials. To accomplish this task the proposal advocates funding for eighty PhD-level geneticists/plant breeders and thirty ecologists.

A modest global effort is already under way. Canada, Australia, and Turkey are working on perennial wheat. In China, a perennial rice is being developed for the uplands. But the effort needs a major boost. So let's imagine that our two countries were to become full partners in this enterprise.

- Russia produced the father of soil science, Vasily Dokuchaev. And soil science will be a must.
- Russia produced the first work on the centers of origin of our major crops under the direction of N. I. Vavilov. As they have over the decades, plant breeders will be going back again and again for germplasm to those centers of origin.
- Ukraine produced the foremost evolutionary biologist of the twentieth century, Theodosius Dobzhansky, who gave us the most important synthesis since Darwin. Ecology and evolutionary biology have become, more or less, one subject since his time, forming a broad discipline now being drawn on by those at work on sustainable food production.
- Researchers and curators who died to protect the seed collection of the Institute of Plant Industry will remain a source of inspiration and hope.
- Russia initiated the first effort to produce a perennial grain.

Though it is not asking for itself any form of official support, The Land Institute will offer to the common effort free germplasm and more than thirty years of experience with perennials. Its staff in this decade has greatly enhanced the diversity of crops and increased the speed of change. We have hybrid prototypes of perennial wheat, sorghum, sunflower, and other crops.

After three decades of collaboration with several land-grant universities and other institutions, we are convinced that we now have a critical mass of researchers who understand the need for systemic change in agriculture. So we in the United States would offer to this effort our trained professionals. We have more PhD-level geneticists, plant breeders, ecologists, and evolutionary biologists than any other country. Our country's

Morrill Act of 1862 made possible our land-grant university system. The Hatch Act of 1887 placed experiment stations in every state. The Extension Service was established by the Smith-Lever Act of 1914. This system worked effectively, but within the context of a country rapidly industrializing by the extravagant use of petroleum and other nonrenewable resources. The now-foreseeable exhaustion of these limited resources is forcing us to adopt a new concept for agriculture. I have spoken here of our reasons to hope that the world's agricultural system can be corrected before it fails.

Finally

Cosmonauts and astronauts, representatives of our two countries, shared a tiny capsule in space. But they brought with them food raised on the soils of our earth.

It seems that we have to be reminded that earth is in space too, as much as are Mars and the moon. Why do we neglect the earth's food-producing system? Why do we allow it to be damaged by erosion and toxic chemicals, all the while depending on fossil fuels, which we are rapidly using up and which will never be cheap? Why do we tolerate this deficit spending?

If we begin now, together, to move agriculture from an extractive to a renewable economy, perhaps fifty years from now at this Coon Rapids, Iowa, farm, those assembled will celebrate the two hundredth anniversary of Darwin's *On the Origin of Species*, the one hundredth anniversary of the friendly meeting between Nikita Khrushchev and Roswell Garst, and the fiftieth anniversary of our effort to end deficit spending of ecological capital. And maybe this day will be seen also as the moment we began to make nature's economy the standard by which to judge our agricultural economy.

After the Talk

After I finished my talk, Sergei Khrushchev, now on the faculty at Brown University, spoke, and then all two hundred of us had lunch on the Garst lawn with the same menu as fifty years earlier. The food was great. It was a beautiful day.

At the end of lunch, Sergei Khrushchev pulled the Russian secretary of agriculture and me aside, and I extended my plea for a cooperative effort between the USDA and the Russians because of their rich history. Through the translator came a response for which I was not prepared. The secretary of agriculture had said, "I need to be blunt with you. We don't have the scientists." I said, "Well yes, I know about the Great Purge in Genetics during the Stalin era, when some three thousand were forced out of genetics work because of Lysenko." "No, no," Khrushchev said, "not Lysenko, Yeltsin. The scientists all left." Sergei was one of them.

We exchanged pleasantries, exchanged cards, shook hands, and went our separate ways. By two in the afternoon Joan and I were driving home, both ruminating on what had transpired, beginning with the reception and dinner the night before.

I thought back to that era following Mikhail Gorbachev with Boris Yeltsin and remembered an Associated Press article I had read in 1997 about a Russian welder in Siberia, "Russian Workers Cope as Best They Can." At home I reread that AP article:

> Each day, Nikolai and Galya arise in the dark and go about the business of making a living. They milk their cows, feed their pig, gather eggs from their chickens, tend their garden. They live off what they grow, and sell the rest for a few rubles here and there. From milk alone, they earn perhaps $100 a month. And when the sun rises, Nikolai heads off from his simple wooden house to his long-time job as a welder in a state-run auto repair factory. For this, he earns nothing.

The article continues, "People survive on their gardens and their wits, and the official economy primarily is a distraction." After some mention of an impending trade union strike and President Boris Yeltsin's concern about doing something about it, the writer says more:

> Across Russia, especially in smaller towns and villages, millions of workers have gone months without wages. Both the government and private employers have been unable—or unwilling—to pay them. Even retirees have gone without their pensions. Outsiders tend to ask how this is possible: How can a nation survive when its people are unpaid? Why would a worker show up for a job that offers no wages? Like many Russians, Nikolai—who hasn't been paid in three months—doesn't ask these questions. Why wouldn't he show up for work? "Where would I go?" he said. "There aren't any other jobs in this town. I'm too old to look for work in Moscow. This is a one-factory town; we have no other choices. And besides, what if the day I decide not to show up the managers start handing out wages?

A message not explicitly mentioned in the article is that nature's economy in combination with traditional culture made this family *resilient*. It continues to feed the people, and it sustained the industrial economy.

Imagine nearly anyone but the Amish going with no wages in the United States for three months now that our traditional rural economies have been mostly undone. The collapse of the Soviet Empire represents the first major failure of the Industrial Mind. Both systems have sought to concentrate power and in so doing have greatly reduced the number of people on the land and in small communities.

What is needed for all of us here, for us, the Russians, and the rest of the world, is to increase our *resilience thinking* while we have slack.

CHAPTER THIRTEEN

WERE ANTS THE FIRST AGRICULTURISTS?

History celebrates the battlefields whereon we meet our death, but scorns to speak of the plowed fields whereby we thrive.

—Henri Fabre

This story is about ants, mentioned here under the cover of whether agriculture arose as a mistake or an accident. To make a mistake means to have chosen wrongly. An accident, on the other hand, is an event or condition occurring by chance. The distinction between mistake or accident may not be valuable, but it is charming enough to be retold. I first read about it in a book by H. E. Jacob entitled *Six Thousand Years of Bread*.

The story goes that less than two years after publication of *On the Origin of Species*, Charles Darwin addressed the Linnean Society of London. It was April 13, 1861, one day after the Confederates fired on Fort Sumter. He was fifty-two, the same age to the day as Abraham Lincoln. While Lincoln was preparing for war, Darwin arose to talk on the origin of agriculture. He began his lecture by telling of two letters from a Dr. Gideon Lincecum, a physician in Texas. Lincecum had written Darwin that agriculture was not invented by man, that ants were the first sowers and reapers of grain.

From Lincecum's first letter regarding ant agriculture, Darwin read as follows:

> Around the mound . . . *the ant clears the ground of all obstructions, levels and smooths the surface* to the distance of three or four feet from the gate of the city, giving the space the appearance of a handsome pavement. Within this paved area not a green thing is allowed to grow, except a single species of grain-bearing grass. Having planted this crop in a circle around the center of the mound, the insect tends and cultivates it with constant care, cutting away all other grasses that may spring up amongst it and all around outside of the farm circle to the extent of one or two feet more. The cultivated grass grows luxuriantly, and produces a heavy crop of small, white, flinty seeds. The insect then harvests the crop, and after the harvest all the chaff is taken out and thrown beyond the limits of the yard area. This ant rice begins to grow in the early days of November. There can be no doubt of the fact that the particular species of grain-bearing grass is intentionally planted.

Darwin answered Dr. Lincecum by expressing his doubts, which prompted a second letter. The Texas doctor emphasized that he had been a student of such ant behavior for the past dozen years and for all seasons had watched this deliberate sowing of a grain monoculture. Darwin then read from the second letter:

> Not a green thing is suffered to grow on the pavement, with the exception of a single species of grain-bearing grass: *Aristida stricta*. . . . They also collect the grain from several other species of grass as well as seed from many kinds of herbaceous plants— but they never sow them. They sow it in time for the autumnal rains. About the first of November a beautiful green row of the ant rice about four inches wide is springing up on the pavement, in a circle of fourteen to fifteen feet in circumference. In the vicinity of this circular row of grass they do not permit a single spire of any other grass or weed to remain a day. Leaving the ant rice untouched until it is ripe, which occurs in June of the next year, they gather the seeds and carry them into the granaries.

Anticipatory behavior is not at all uncommon among species other than humans. The showstopper here was that the ants had sown the grain. This made them farmers. The lecture hall was alive with questions. Some wondered if Lincecum was putting them on or was simply in error. H. E. Jacob in his book *Six Thousand Years of Bread* takes up this question and concludes that it was finally answered by Ferdinand Goetsch in 1937. Jacob writes:

> Lincecum's ants were not fabulous, they had "actually sowed by mistake." From extensive observation of ants in Europe and South America, Goetsch discovered that most ant tribes are possessed of two contradictory instincts: the instinct to collect and the instinct to build. The first leads them to bring grain, wood and other materials into the nest; the other instinct impels them to carry all these materials out of the nest and use them for building. Among their building materials are seeds of grasses and grains. A period of drought stimulates the collecting instinct; moisture, the building instinct. Goetsch suggests that Lincecum's fields arose from such grains which were aimlessly (for they should have

been used for food) carried out in the course of building activities. Thus they were unintentionally placed upon moist soil and germinated. Conscious sowing it was not. But if this *"mistake"* had been repeated millions of times in the history of the race, thereby becoming a part of the group memory, could not this in the end engender a sowing instinct? Perhaps Lincecum's particular ants had developed such an instinct.

I emphasize the word "mistake." At The Land Institute we have concluded that human agriculture arose as an "accident." We have continuing conversations about the origin of agriculture because of the very practical need for us to gain insights for those of us devoted to perennializing the major crops. A question always before us is why the first human agriculturists selected the annual species instead of perennials for grain production. Our senior scientist and geneticist Stan Cox has described what he calls "the contingent nature of annual crop domestication":

> Agriculture was conceived in probably three locations on the planet (the Fertile Crescent of southwest Asia, Central America and maybe China). Farming began in the first and most influential area of origin, the Fertile Crescent, soon after the retreat of the last Ice Age.

Whyte (1977) noted that the rapid warming of the earth's climate at that time created three large and three small "arid cores" on the Asian continent. On the fringes of these cores there formed concentric "isoxethermic zones," where peak temperatures occurred during the dry season.

Annual grasses and legumes were the first and most prominent plants to thrive under the new climatic conditions. They largely displaced perennial species in the Fertile Crescent, becoming "suddenly and abundantly available" to human hunter-gatherers. The sudden, wide availability of large-seeded annuals attracted humans, who previously had relied in part on seeds of perennial grasses for food. They began to harvest and sow, initiating the agricultural revolution and never looking back.

As Jared Diamond argues in *Guns, Germs, and Steel*, the idea of (annual) agriculture spread in a wave that preceded crop development. As people moved across the continents, they carried seeds with them, and when they reached areas where their crops were no longer adapted, they, or someone else, either selected the existing crop species for better adaptation or domesticated local wild annuals, following the by-then-familiar pattern of sowing a part of the harvested grain.

In this way, annual cropping moved far beyond the Mediterranean climate where it originated, into almost every soil and climatic zone on the planet. Early in that process, humans became dependent on the annual crops and reached an apparent point of no return. But, ironically, an early product of annual agriculture—civilization—has developed to a point at which we now can see—ten thousand years later—that annual agriculture was largely a historical accident. We at The Land Institute maintain that it is a reversible accident.

Lincecum's "mistake" and Stan Cox's "accident" converge here to become near-synonyms. For ten thousand years, humans have been beneficiaries of this "accident." Had the ants achieved a scale equal to ours and featured annuals, it is likely that they, too, would have sent the natural fertility seaward. Our farming has never been sustainable. For ten thousand years humans have expanded agriculture into new land, even when population growth did not warrant it. These ventures have been seen as virtues. We speak approvingly of "groundbreaking research" or "plowing new ground." That agriculture is a way to waste or mine the soil is seldom mentioned. In retrospect, a faint foreshadowing of the fossil fuel epoch can be seen in our agricultural beginnings. That this inherent wastefulness of the source of our sustenance and health happened at such a slow rate, seldom perceived by most people in one generation, is beside the point. The fossil fuel epoch, with its subsidy of fossil carbon into agriculture, which sponsors other nutrients, obfuscates for most of us today how little natural capital is available in the soils of the globe.

PART IV

Analyzing the Resistance

CHAPTER FOURTEEN

ANALYZING THE RESISTANCE

.

NOTHING FAILS LIKE SUCCESS

Green Revolution yield increases were two-and-a-half-fold from 1960 to 2000. The negative consequences were ecological, genetic, cultural, and in social justice. These four were more or less ignored.

Part of the problem has to do with the history of recent successes in agriculture where the primary features were bushels and acres. Yields increased dramatically. Some consequences were unforeseen, some were unforeseeable. (There is always that invisible line.) The ecological consequences included soil erosion in both quantity and quality. The same with water. Countless land races disappeared. Therefore, genetic diversity was lost. As large scale became more widespread, the eyes-to-acres ratio declined. Fossil fuel inputs reduced the need for traditional practices for pest and fertility management. Finally, as economies associated with small scale declined, people moved to cities. Industrial—which is to say fossil-fuel-driven—minds and associated capital dictated the patterns on the landscape.

There were gains, such as the two-and-a-half-fold increase in food production. It is worth remembering, however, that the genetic "improvement," together with more fertilizer, more pesticides, expanded irrigation, farm equipment modifications (to accommodate scale and fossil energy use), all combined with luck to find a few useful genes.

The message to the Industrial Mind was clear. A problem was identified. A narrow mission was established. The capital to achieve the goals of the mission was acquired. Yields increased. Negative consequences can be more or less ignored.

IS IT THE NATURE OF OUR BRAINS?

The late Harvard biologist Ernst Mayr filled nearly one thousand pages outlining the growth of biological thought. Mayr devotes several paragraphs to great men in science, noting that the greatest ones "have modified their ideas most frequently and most profoundly." He adds that "scientists who never changed their major ideas from the beginning to the end of their career are probably a small minority." But the greatest changed their major ideas, most often profoundly so.

Of course, scientists are not alone in their ability to accept new evidence to make connections. Lots of disciplines promote objectivity, and some people just have it. But what about those among us who are *not* inclined to be open-minded? That is a challenge for all of us.

Experimental neurologists have observed growth in the neurological network. Rats that learn actually have growth of dendrites, nerve endings, into parts of the brain. If the nerve network in the brain is not there, they can't perform their task.

So, here's the question: Is the reason so many members of the society are so slow in responding to the problems of our ecosphere that they lack the neurological network, those dendrites that flare out at the end of the nerve?

For example, what causes so many of our citizens to discount the consensus of our own National Academy of Sciences and the consensus of the Intergovernmental Panel on Climate Change? They seem not to respect the scientific way of knowing. Why the denial? Scientists are often wrong, but it is other scientists who are also almost always the corrective agents. Here we have a consensus of two thousand scientists from around the world.

Perhaps there is something about the nature of our brain that is limiting. A *Washington Post* article (September 4, 2007) described how the federal Centers for Disease Control and Prevention had issued a flier to combat falsehoods about a flu vaccine. The flier recited various commonly held views and labeled them either "true" or "false." Among those the CDC identified as false were statements such as "The side effects are worse than the flu" and "Only older people need flu vaccine."

A University of Michigan social psychologist and researcher saw the flier and had volunteers read it. His findings were astonishing. Within *thirty minutes* after reading, older people misremembered 28 percent of the *false* statements as *true*. Three days later, their memories now had it at 40 percent factual, up from 28 percent. Being incorrect increased with time.

Younger people did better at first, but three days later they were as wrong as older people were after thirty minutes. Moreover, both the young and old believed that the source of their false beliefs *was* the respected CDC.

This has implications for public policy. When we try to correct a falsehood or urban legend with accurate information, the studies show that denials and clarifications *contribute* to the popular falsehoods. For example, large numbers of Americans incorrectly believe that Saddam Hussein was directly involved in planning the September 11 attacks, and that most of the hijackers were Iraqi. The experiments suggest that both "intelligence reports and other efforts to debunk this account may in fact help keep it alive."

Many Arabs "are convinced that the destruction of the World Trade Center was not the work of Arab terrorists but was a controlled demolition; that 4,000 Jews working there had been warned to stay home that day; and that the Pentagon was struck by a missile rather than a plane." Years ago it was commonly said that people believe what they want to believe.

In 2006 the Pew Global Attitudes Project "found that the number of Muslims worldwide who do not believe that Arabs carried out the September 11 attacks is soaring—to 59 percent of Turks and Egyptians, 65 percent of Indonesians, 53 percent of Jordanians, 41 percent of Pakistanis, and even 56 percent of British Muslims."

With climate change, it is the problem of trying to fight bad information with good information. Clever manipulators can take advantage of the weakness just described.

We know that not everyone believes the falsehoods. *The point is, the mind's bias affects many people.* Once an idea is in there, it can be hard to get out. If you repeat that the information is wrong, you may be reinforcing the wrong information, making repetition a key culprit. In politics and elsewhere, *whoever makes the first assertion has a great advantage over later deniers.*

Accusations or assertions met with silence *are more likely to feel true,* according to another study.

A SOCIETAL PROBLEM?

Beyond the individual, the *problem of society* at large is daunting. Morris Berman (*Dark Ages America: The Final Phase of Empire*, published in 2006) believes that every civilization is a "package deal" and that the configuration of that package, more or less, dictates that there is a trajectory imposed by the constraints of that "deal," some positive, others negative. His important idea is that they crystallize into a specific pattern or direction early.

Berman's idea is that the factors that made our civilization's rise to power possible are the "factors that could do it in." For us to be where we are today required that we reject countless alternatives along the way. It reminds me of plant breeding. Collapse or decline can only be avoided if the *alternatives* that have been repressed are incorporated in the dominant way of being. Plant breeders keep various breed lines alive in case they are needed later.

Corporations have short-term interests, and their managers place foremost their fiduciary responsibility to stockholders. Meddling politicians are a liability.

In the July 23, 2007, *New Yorker* (page 25), we read that "in the auto industry, there's one thing you can always count on: if a new environmental or safety rule is proposed, executives will prophesy disaster. In the nineteen-twenties, Alfred Sloan, the president of General Motors, insisted that the company could not make windshields with safety glass because doing so would harm the bottom line. In the '50s, auto executives told Congress that making seat belts compulsory would slash industry profits. When air bags came along, Lee Iacocca told Richard Nixon that 'safety has really killed all our business.' A few years later, when Congress was thinking about requiring fuel-economy standards, auto executives warned that instituting such standards would create 'massive financial and unemployment problems.'"

They have been effective. For twenty years, fuel economy standards were not raised.

A PROBLEM OF CULTURAL AND REGIONAL HISTORY?

There is the problem of our history and how we came to have the mindsets that we have. I live at the boundary of the Midwest and Great Plains, perhaps the place with the most blatant examples of having to rely on a series of mining economies. First came the white hide hunters, until the bison were finished off in the 1870s. Then came the miners of bones, which were shipped east. Then the miners of grass, some might say. Followed by oil miners and gas miners. Then the soil miners, which began with the great plowing of the 1910s and '20s followed by the dust bowl. Then the miners of water over the Ogallala Aquifer.

More stands behind that series of mining economies. The late Dan Luten, in the Geography Department of University of California–Berkeley, professor and friend, described it well three or four decades ago: "We came as a *poor people* to a *seemingly empty land* that was *rich* in resources, and we built our institutions for that *perception* of reality: poor people, empty land, rich. These perceptions helped shape our social, political, economic, and even religious institutions. Some called it our Manifest Destiny. The nature of the reward all predicated on support from all of those institutions. Poor people, empty land, rich."

As a corollary of sorts, we have Wendell Berry's insight: "We came with vision [to this continent], but not with sight. We came with visions of

213

former places, but not the sight to see where we are." He tells how we came as both settlers and conquerors and found natives. Soon those natives became surplus people standing in the way of civilization, which required settlement, which is to say, settlers. These natives, now regarded as surplus people standing in the way of settlement, became redskins, and now descendants of those settlers, nearly everywhere in rural areas become surplus people. They are the new redskins. The mind-sets and institutions that validated the surplus-people concept, redskin people, is still in place in our time.

Lest we lay the blame at the feet of our country's origins only, it is worth considering how many of our other assumptions were hatched in Europe long before we arrived in the early 1600s. The early thinkers of the Enlightenment: Copernicus in Poland, Francis Bacon in England, René Descartes in France, Galileo in Italy (who died in 1642, the year Newton was born). And later John Locke, and Adam Smith, who wrote *The Wealth of Nations* in 1776. Benjamin Franklin and Thomas Jefferson, plus John Adams and the other signers of the Declaration of Independence, were influenced by the Enlightenment. Their devotion to those ideals was partly what stood behind their publicly signing their own death warrant. The language of that document carried the most advanced thoughts of the time.

We are all beneficiaries of that document today, proudly so, I might add, but its promises being fulfilled are in the mix of what we must think through in our time.

On such a historical journey, one doesn't know where to stop. Abraham probably had a hand in it. Moses led a journey that was clearly an example of Manifest Destiny. The ancient Greeks, Plato, Aristotle, the caesars in Rome. The Visigoths were there. Charlemagne watched the clearing of the forests of ancient Gaul. They all sent their ideas into the

Dark Ages and the feudal system, the Renaissance, the Reformation, and beyond. But it was early in the seventeenth century, the time of Bacon and Descartes, when their ideas began to be carefully codified. And now, "perfect memory," as George Bernard Shaw said, "is perfect forgetfulness."

North America was regarded as a clean slate, and the founders of our country, as children of the Enlightenment, wrote the code for our behavior. That is our "package deal" that Berman was referring to, and it is important to remember what he said: collapse or decline can only be avoided if the *alternatives* that have been repressed are incorporated in the dominant way of being. When resistance to the alternatives is fierce, then we are destined to get where we are headed.

PART V

Away from the Extractive Economy

Chapter Fifteen

AWAY FROM THE EXTRACTIVE ECONOMY

We are vastly superior to any other species in stretching our world into the shape we want; that also makes us infinitely more capable of creating unforeseen difficulties. As a general rule, the greater the changes we think into being, the more problems we have to face. Environmental history is, among other things, a lengthy account of human beings over and over imagining their way into a serious pickle.

—Elliott West, *The Contested Plains*

A t the species level, *Homo sapiens* has demonstrated no sign that we have the ability to practice restraint. I have been relentless thus far to emphasize that like bacteria on a petri dish, fruit flies in a flask, or rabbits without predators, if the energy is there, our numbers increase. As powerful creators of abstractions, we have developed an economic system that, as it stands, moves our population and the spending of materials and energy as fast as possible to the edge of the petri dish. Having been effective at death control through diet and medicine, we are now forced to exercise restraint. Where do we turn?

We have a few examples of people whose lives serve as examples of alternative ways of living, and we can imagine that one day a Chamber of Resilience replacing the Chamber of Commerce across our land.

EXHIBIT A: LELAND LORENZEN, ADOPTER OF VOLUNTARY POVERTY

My friend Leland Lorenzen died September 6, 2005, not far from his seventy-ninth birthday. He had lived on less than $500 a year for nearly all of twenty-nine years in a shack six by sixteen feet. A small wood-burning

Figure 23. Leland at his shack (photo by Terry Evans)

stove provided heat. He ate mostly soaked wheat, greens from his yard, and from time to time, milk from his goat. He died as he had lived. A day or two before he died he turned to his son, Jule, and said, "Time to open a hole." The family buried him in his sleeping bag on his one-acre plot, his grave dug with a fossil-fuel-powered backhoe by a neighbor who refused to accept pay for the digging. His burial did not cost the family one cent.

It has been my experience that those who come the closest to "walking the talk" don't actually talk about it. Leland was the best example I know of a "walker," but he too was dependent on the extractive economy, including for his burial. Even Leland was never off the grid. He acknowledged that he was grid-dependent, in fact, especially when he was a merchant seaman. It was being a merchant seaman for seven years—he had been around the world in various ports—that contributed to his radical views on economies. In ports near and far he noticed the difference between the economy of the street and the economy of the official culture. But while he was working at the local oil refinery he read Thoreau's *Walden*. What he had been turning over in his mind came together. He threw a copy of the book on the kitchen table and told Bernice, his wife, she had better read it because it was going to change the life of their family, which included three children, ages four, eleven, and thirteen. During this period he was quite apocalyptic, certain that nuclear war was inevitable.

Beyond *Walden*, one of Leland's insights was that we start doing violence to people and the environment when we seek pleasure. He insisted that there is nothing wrong with the experience of pleasure, but when we *seek* it we start manipulating the world, people included. Conflicts between individuals or nations come from the same source, he often said. Here is how it works. In our heads we have an imagined environment that will bring us pleasure, and through the pursuit of pleasure we begin to try to duplicate it in the world others have to live in with their own imagined environment. This leads to conflict. The pursuit of making the environment real, and of that pleasure, is in conflict with the imagined environment and pleasure other people are pursuing. In such a manner we destroy the world's fabric. To most people, I suppose Leland carried his beliefs to an extreme. When I first knew him, he had a beautiful garden and in the winter a hotbed of sorts that extended the growing season of various vegetables. Then one year he quit gardening, reckoning that that too was a form of pleasure seeking. From then on it was greens out of the yard, soaked wheat, and milk from his goat.

He shortened a rusted-out Volkswagen Kharmann Ghia convertible and drove it to conserve fuel. It looked like Donald Duck's car. During his tenure in the shack he figured he spent $350 a year on gasoline and tobacco and $150 on food.

Leland lived thirty miles south of The Land Institute, relatively isolated in the country. Late one winter when I was on my way to the Wichita airport to catch a plane, I left enough time to stop by his shack to see how he was and to visit. I noticed that he had been poking seeds into a flat of dirt. I asked Leland, "What's going on here? You're going to have a garden again. You're back into pleasure seeking." "Come on in, Wes," he said. "I'll tell you all about it. I'm all [screwed] up." According to Leland, Bernice, who lived in a bomb-shelter-type house some seventy feet away, felt she needed $300 a month to meet expenses. Her Social Security would only allow her $200, but because Leland had worked at regular jobs before reading Thoreau, he was eligible for $400 a month. So Leland took out Social Security and was giving Bernice $100 per month. The remaining $300 was, as he put it, "piling up in the bank." That accumulation was causing him to "have creative thoughts," and he had started various projects in his shop and on his place. Indeed these thoughts led him into pleasure seeking again. His brain was "on fire with imagination." He began to imagine pleasure foods he knew he could do without.

I couldn't help him out of his dilemma, and besides, I had to get to the airport and contribute to more atmospheric carbon. This was March. Leland continued to draw on his Social Security until around Christmas. My daughter Laura had come home for the holidays. On a cold Christmas Day, Laura and I drove toward Leland's with the idea we would visit the Maxwell Game Preserve to check on the herd of bison and prairie elk, have a winter prairie picnic there, and then go see Leland. We parked our car at the preserve and walked a couple of miles through pasture, fields, and snow to Leland's. He was in his shack, and once again he began to complain about the money piling up in the bank forcing him into creative thinking. Laura said, "Well, Leland, why don't you just quit taking that Social Security since it is bringing you so many problems?" A month or so later, Leland drove up to tell me, "Laura's words kept ringing in my ears, and I'm going to scratch my name off the Social Security list."

He told me later it wasn't as easy as he thought. He tried, but the official told him that if he did, he would have to give back all that he had received. Well, of course Bernice had spent her allotment and Leland had spent some on his "creative efforts." Leland said he couldn't give it back. The official told him to give it to a worthy cause. "There are no worthy causes," he said. "Give it to your children," the official suggested. "They aren't worthy either," he replied. (I think he thought it would be a source of problems for them, because I know he loved them dearly.) He was stuck. The money had become a curse.

This idea of money as a curse reminds me of another incident involving Leland more than a decade earlier. I was putting a roof on what became the new classroom at The Land Institute after the other had burned down.

Leland stopped by and, crawling up the ladder, began to help. It was inexpensive roll roofing. He helped me immensely that day, and I had both a twenty-dollar bill and a five in my billfold. I tried to give Leland the twenty. He refused. I then tried to give him the five. He said no. I insisted, saying that I could afford it and that he was to take it. The conversation ended when he said, "Don't give me your problems."

Countless numbers of people have asked me what he did with his time. I do know that he had an intellectual life. He would go to the public library, check out six books, read them in the shack, return them, and get six more. He had read a lot of Krishnamurti and knew the Bible pretty well. Though he had serious doubts about an afterlife, he liked a lot of what Jesus had to say and had a particular fascination for the old Jews in the Hebrew Scriptures. He once said to me, "When you have Jews come around The Land Institute, I want to meet them if they take their religion seriously, especially the Orthodox Jews."

A major effort for Leland was to stop the internal dialogue. "We are always either building or protecting an image," he told me. He also thought it nearly impossible to rid the mind's desire to do so in the presence of another person. This was a source of worry. He said the only way out of it was to be alone, and after a while his image of himself would fade and then he would have the "awareness of a squirrel." A squirrel's awareness is of the "effective," immediate environment surrounding him. One can then be out of the environment where the buds of violence would grow.

He took me once a couple of miles from his shack, to an abandoned pasture with prairie and trees all around where there were some large protruding rocks. Using those rocks and a minimum of building materials, Leland had recently built a small shelter just large enough to sleep in. It was really spare. The pillowcase was stuffed with prairie grasses. Here the deer and other wild animals would come to lie down outside, within a few feet of where he was sitting or sleeping, undisturbed by his presence. As I surveyed the surroundings and inspected his handiwork, he explained, "Here is where Leland goes to get away from Leland." I didn't ask what he meant and still think it would have been improper to do so, but I have wondered. My first question was, "Why isn't life back at his shack enough?" I thought of Francis of Assisi, some of the mystics, some monks, Elijah, and other examples. Perhaps even alone an image of who we are begins to form.

Once when I brought a "certifiable intellectual" to visit with Leland, Certifiable Intellectual was disappointed. He said I had given Leland too much credit and thought that Leland had nothing to offer. I didn't argue. It was good enough for me that Leland had become one of my indispensable friends.

He once told me that the wheat he ate daily cost him about 3 cents and that when he had the goat for his milk, the wheat that went into that goat cost 9 cents. The value of the wheat came to less than $45 per year. Now and then he would mix honey into his soaked wheat. He called honey a pleasure food that he could do without. His method of fixing the wheat began by measuring out what he would consume the next day. He would soak the wheat in water and cook it with a forty-watt lightbulb a few hours. (Bernice's house has electricity.) The cover for his container was a hubcap.

Once we took a driving trip in a pickup truck to Long Island, New York, to bring home tools and equipment that had been donated to The Land Institute by a friend whose husband had died. On this trip Leland declared he wasn't going to let me eat him "under the table." But I did. He got so sick we had to stop at a grocery store and buy him some peanut butter and crackers. Back to a simpler diet, he was fine. On another trip one January, he took me around to visit some dropout communities in the Missouri Ozarks.

Most of the people had been antiwar protestors, civil rights activists, and the like, who had thrown themselves out on the land, taking their advanced degrees from places like University of California–Berkeley with them. It was near-subsistence living. It was a good trip, and I enjoyed the conversation. Leland and I talked about that trip many times, noting how difficult it is to try to become a satellite of sustainability orbiting the extractive economy. Over the years most of these idealistic, strong individuals found themselves increasingly pulled into the orbit of the dominant culture.

Once Leland told me that his moments of depression, which sometimes lasted a few days, came as a gift of sorts. He said that it was like a "great cleansing," that it was followed by great clarity and insight. Like Thoreau, he had many visitors who were attracted by his philosophies. Several months after his death, The Land Institute purchased the shack from Bernice, moved it thirty miles north to The Land Institute, and refurbished it a bit. It now stands as a monument to the most bottom-line person I have ever known. The visitors today can see taped to the wall of Leland's shack a government form. He had initialed each "I DO NOT WANT: surgery, heart-lung resuscitation (CPR), antibiotics, dialysis, mechanical ventilator, tube feedings." Then he wrote "or any other health care treatment. I wish to heal or die naturally. Please take me to my bed." *Please* is underlined three times and *my* is circled. He had two witnesses. Ten years and three months later, he died. He was buried the same day on his one acre.

EXHIBIT B: UNCLE JOHN

Uncle John's wife, Aunt Minnie, had died. Aunt Minnie was my mother's aunt. She and Uncle John had one child, a son named Luther. Luther, married only briefly, had also died. If my memory is correct, when Aunt Minnie died Luther's wife got the farm, which she immediately sold. Uncle

John had little money and no place to stay, so he came to live with my Aunt Ruth, my mother's younger sister. Uncle John had what he called a game leg, with an open sore that never healed. He had an upstairs bedroom in Aunt Ruth's big house and took his meals with her and two of her kids, my cousins Danny and Martha.

Aunt Ruth lived on the farm next to ours, on the farmstead of my mother's parents. Aunt Ruth's husband, Uncle Art, had died during the war—not in the war, but during. Their oldest son was fighting the Japanese in the Pacific.

I would go over there in the evenings to play board games and cards with Uncle John and whoever else was interested. Mostly it was Uncle John and me. When bedtime came, Uncle John would hobble up the steps and Aunt Ruth would call after him, "Uncle John, did you apply the salve and dress your leg?" He would call back, "Thank you, Ruth dear, it is taken care of." He had not lied, for I had seen Uncle John expose his sore to the dog, who dutifully got up, walked over, and licked the wound. This, apparently, was all the treatment that Uncle John thought the leg needed. I have had people learned in the art of healing describe the merits of the dog treatment. To this day I have no idea what to make of it.

What brings Uncle John up is not the leg or the dog, or the loss of his farm and his dear wife, Minnie, and his beloved son, Luther, though volumes of tears were shed in that kitchen by the cookstove (if it was winter) over those losses. What I want to talk about is the game of Monopoly, which the two of us frequently played.

I won every time, and not, I think, because I was a child and he let me. Every time it would be the same story. We made the rounds as dictated by the roll of the dice. Uncle John might land on Park Place. He would look at the price, count his money, roll his cigar from one side of his mouth to the

226

other, and always decline. "That's a little steep for me," he'd say. But when Uncle John landed on a low-priced property, Baltic for example, if he had the money he might snap it up.

You can appreciate where this story is headed. I ended up with the expensive properties. He ended up with the cheap ones. I put up houses and hotels every chance I got. He might sprinkle some houses on those cheap properties, but would decline the opportunity as often as not. It wouldn't be long before he landed on one of my properties, often with one or more houses or a hotel. When he heard the rental price, it hit him like a force. If he was low on cash, he would ask, "Can you hold off until I pass Go, Sharon?" (That is my first name, which I went by until college.)

At first I would hold off. But then he'd hit another one of my expensive properties and I'd suggest that he sell a house or mortgage one of his properties.

"Naw, I ain't gonna mortgage," he would insist. "I saw what happened to my neighbors in the Depression. I vowed never to mortgage my place, and I kept it—me and Minnie. We just cut back." And then he would say, "I wish I had that place now. I could still make a go of it, even with my game leg, with a few chickens and my garden."

Well, that's how I beat him every time. I beat him because the game is rigged. I beat him because he was from the old school, which operated, to quote Milton, "according to the holy dictate of spare temperance."

EXHIBIT C: THE WOMEN OF MATFIELD GREEN

In the early 1990s I lived on and off in the small Kansas town of Matfield Green, population fifty-six. The Land Institute had a presence there, and I acquired a few abandoned houses. At work on them, I had great fun tearing off the porches and cleaning up the yards. But it was sad, as well, going through the abandoned belongings of families who lived out their lives in this beautiful, well-watered, fertile setting. In an upstairs bedroom, I came across a dusty, but beautiful, blue padded box labeled "Old Programs— New Century Club." Most of the programs from 1923 to 1964 were there. Each listed the officers, the Club Flower (sweet pea), the Club Colors (pink and white), and the Club Motto ("Just Be Glad"). The programs for each year were gathered under one cover and nearly always dedicated to some local woman who was special in some way.

Each month the women were to comment on such subjects as canning, jokes, memory gems, a magazine article, guest poems, flower culture, misused words, birds, and so on. The May 1936 program was a debate: "Resolved that movies are detrimental to the young generation." The June roll call in 1929 was "The Disease I Fear Most." That was eleven years after the great flu epidemic. Children were still dying in those days of diphtheria, whooping cough, scarlet fever, pneumonia. The August program was dedicated to coping with the heat. Roll call was "Hot Weather Drinks"; next came "Suggestions for Hot Weather Lunches"; a Mrs. Rogler offered "Ways of Keeping Cool."

On August 20, the roll call question was "What do you consider the most essential to good citizenship?" In September of that year it was "Birds of our country." The program was on the mourning dove.

What became of it all?

From 1923 to 1930 the program covers were beautiful, done at a print shop. From 1930 until 1937, the effects of the Depression are apparent: programs were either typed or mimeographed and had no cover. The programs for two years are now missing. In 1940, the covers reappeared, this time typed on construction paper. The print shop printing never came back.

228

Figure 24. Matfield Green, Kansas, women, the New Century Club, late 1930s or 1940s

The last program from the box dates from 1964. I don't know the last year Mrs. Florence Johnson attended the club. I do know that Mrs. Johnson and her husband, Turk, celebrated their fiftieth wedding anniversary, for in the same box are some beautiful white fiftieth-anniversary napkins with golden bells and the names Florence and Turk between the years 1920 and 1970. A neighbor told me that Mrs. Johnson died in 1981. The high school had closed in 1967. The lumberyard and hardware store closed about the same time, but no one knows when for sure. The last gas station went after that.

Back to those programs. The motto never changed. The sweet pea kept its standing. So did the pink and white club colors. The club collect persisted month after month, year after year.

A Collect for Club Women

Keep us, O God, from pettiness; Let us be large in thought, in
    word, in deed.
Let us be done with fault-finding and leave off self-seeking.
May we put away all pretense and meet each other face to face,
    without self-pity and without prejudice.
May we never be hasty in judgment and always generous.
Let us take time for all things; make us grow calm, serene, gentle.
Teach us to put into action our better impulses; straightforward
    and unafraid.

Grant that we may realize it is the little things that create differences; that in the big things of life we are as one.
And may we strive to touch and to know the great common woman's
heart of us all, and oh, Lord God, let us not forget to be kind.
—Mary Stewart

By modern standards, these people were poor. There was a kind of
naïveté among these relatively unschooled women. Some of their poetry
was not good. Some of their ideas about the way the world works seem silly.
Some of their club programs don't sound very interesting. Some sound
tedious. But their monthly agendas were filled with decency, with efforts to
learn about a wide variety of items: birds, our government, coping with their
problems, the weather, and diseases. Here is the irony: they were living up to
a far broader spectrum of their potential than most of us do today!

I am not suggesting that we go back to 1923 or even to 1964. But I will
say that those women were further along in the necessary journey to become
native to their places, even as they were losing ground, than we are.

EXHIBIT D: BRIAN DONAHUE AND
WESTON, MASSACHUSETTS

Smart children of parents who have spent much of their productive lives
worrying about the global problems of population growth, resource depletion, and pollution of nature's sinks and sources have been known to ask
questions like this: "And so, Dr. Doom, what can we do about it where *we*
live?" How *can* a suburban Sierra Club member who frets over spotted
owls and pesticides act beyond writing his or her congressperson about the
owls and avoiding the use of pesticides on the lawn and garden? A primary
source for proposed solutions to these problems must be those who have
had a serious engagement with the land itself. Brian Donahue lives in the

Boston suburb of Weston, Massachusetts. He is an environmental historian who teaches at Brandeis University. Donahue's writings bring social and natural history together, subjects that are usually treated separately. Suburbia, we all know, is the product of a prime-farmland-gobbling, mall-producing, loneliness-generating, xenophobic, consumptive life in which secular materialism is now the national religion. The majority of Americans are either already in suburbia or headed there. Suburbia defines our being, the world that even the nonsuburbanite is pulled into. It is the world from which investors in economic growth extract the sort of wealth that dooms our culture and forces our descendants to pay the bills.

Rural landscapes and their declining small towns—and for that matter burned-out inner cities—present problems obvious to any who care to look. The problem is beyond soil erosion and farmstead buildings collapsing, greater than rural poverty and rural drug addiction. Here the Jeffersonian dream of small towns and rural communities has failed.

In suburbia, on the other hand, children of the affluent are raised with the idea that they are in the midst of a legitimate American dream. In Brian Donahue's Weston, most of the kids are fit, play soccer, do well in a first-rate school system, and perhaps after a minor bout of adolescent alienation go on to live affluent lives, fretting a little now and then about the loss of rain forest. Sure, suburban kids might be dreadfully overweight, TV-watching and arcade-playing, attending schools that reward minimal compliance, but whatever the situation, most failures within the suburban landscape are due to economic success. The suburban dream rests on the social and environmental nightmare that haunts the inner cities and the harrowed countryside.

Weston, Massachusetts, is a suburb. Why would an environmentalist be drawn to a place like Weston, where the inhabitants perceive that life is as good as it gets? Donahue became interested for that very reason; he saw that what suburbia needs is not bumper-sticker or T-shirt environmentalism. It is not enough to sport an artfully silk-screened "Think Globally—Act Locally" T-shirt and limit that responsibility to recycling plastic in six categories.

Most people *need* a landscape, and essentially all agricultural landscapes in the United States *need* more people. What about the fact that most suburban or urban kids want to work in a meaningful way but cannot do so? Nearly twenty years ago the cultural anthropologist Mary Catherine Bateson blurted out in a conversation, "Boredom has to be taught." We teach boredom to our suburban children in countless ways. Children need reality, not virtual reality. Children need to understand our source for survival, whether it is wood for a stove or food for their bellies. Donahue's efforts to show connections between land and food to the young will relegate the environmental bumper sticker and T-shirt to some future museum dedicated to environmental and social naïveté. Brian Donahue has helped create a new genre, a success story out of the suburbs sure to give heart to sociologist and ecologist alike.

If America doesn't save agriculture, wilderness is doomed. It is doomed because hungry people will encroach upon it and destroy it to meet immediate food needs. What better place to teach the essentials of agriculture and forestry than in the suburbs, where most people live? That is what Professor Brian Donahue has been about for over three decades.

EXHIBIT E: RESILIENCE VERSUS ECONOMIC GROWTH: THE NEED FOR A CHAMBER OF RESILIENCE

Parochialism and provincialism are direct opposites. The provincial has no mind of his own; he does not trust what his eyes see

until he has heard what the metropolics . . . has to say. . . . The parochial mentality on the other hand is never in any doubt about the social and artistic validity of his parish.
—Patrick Kavanagh, Irish poet

Imagine that a few of us in our various regions chose to exercise our imagination along the lines that follow because we perceive it necessary to measure our progress by how independent of the extractive economy our region can become. Were our ideas taken seriously, we would begin to feature resilience over economic growth. I realize that is unlikely to happen very soon, for imagine a Chamber of Resilience rather than a Chamber of Commerce. It would be fruitless to ask the Chamber of Commerce to disband, given that we are all embedded in too many centuries of economic history in civilization. But the time will come in the foreseeable future when the natural limits to growth of population and resource throughput will impose some cruel consequences upon us. But what if we could imagine setting some realistic mileposts toward a more renewable, less vulnerable future? Difficult? Yes! Impossible? No. Unlikely? Probably, precisely because people of the world almost everywhere now have been beneficiaries of mining economies stretching back ten to twelve thousand years. It is everywhere in our country, of course, but perhaps most notably in our Great Plains region, where extraction has been the way of life for humans since the arrival of the Europeans, most especially after Kansas statehood in 1861. First it was bison hides, then their bones, oil, natural gas, soil, and water, pretty much in that order. We have seldom asked, as our minds wander, anticipating the end of these extractive economies: "Then what?"

A rising interest in resilience has developed over the last year or so, most especially, I suspect, due to the implications of rapid climate change and oil prices climbing to $100 a barrel. Rather than our continuing to wait until we get closer to the end of the fossil fuels or water, the described reality of climate change has our minds concentrated on the necessity to practice restraint now in the use of fossil carbon, forest products, and soils. Restraint causes us to question economic growth as we have known it, at least for the past 250 years, since the start of the industrial revolution in England.

The Irish poet's distinction seems useful in helping us adopt the mentality closer to the parochial and helping us stand up to the stares of the provincials. The time will come when the abstractions that govern our modern lives—such as what constitutes real wealth—will be shown as hollow. As that hollowness becomes clear to more people, our economy will be more directly founded on our land, our water, the sun, and the ancient virtues of thrift and skill. Our collective ability to look after one another in our community will be on the line. We can set some mileposts now and begin to measure community progress toward our goal of resilience. Satisfaction and even pleasure as well as pain avoided can be imagined. So when down-powering is thrust upon the world, our infrastructure will be in place and we will be better prepared to handle it gracefully.

Community decisions based on resilience would feature keeping our options open, which would require viewing events in a regional as well as a local context. We would feature recognition of our ignorance over knowledge precisely because there will be more future events of the unexpected than the expected variety. In other words, a resilience strategy would avoid the planning of systems heavily dependent on knowledge.

With all of this in mind, for starters we would stop essentially all additions to the human-built environment on valley land, except within current city limits. This is only prudent, since food does not come from a grocery store, but rather from ecological capital whose main features now are soil, oil, and natural gas. Withdraw the oil used for tillage of various sorts and the natural gas that serves as the feedstock for nitrogen fertilizer, and what remains is the soil's biological capacity and nutrient supply to produce food. On a global basis, there would be widespread famine. This capital behind agriculture without fossil input subsidies is significantly higher on our valley soils than in the hills.

Two lifetimes ago, my town of Salina, Kansas, began. One lifetime ago the fossil carbon responsible for more than doubling food production was in the early stages of introduction to our fields. Where will we be one and two lifetimes from now? The world population has tripled in one long lifetime, and it is sobering to contemplate that roughly two-thirds of the world population is here because of the sponsorship of carbon and nitrogen out of the oil and natural gas wells.

To cover prime land to accommodate carbon combustion (automobiles, restaurants, motels, building supply places, industrial fabricators, etc.) rather than carbon fixation (wheat, sorghum, soybean fields, and more) is to ignore the fundamental laws of ecological economics. The total regenerative capacity of the region is reduced with every human-built accoutrement of civilization. We can't escape that. That regeneration capacity is reduced the most when placed on valley land. But there is more to the story, much more. If we can keep ourselves fed, at least we will have the option to work through the other challenges sure to come.

PART VI

An Enigma

enigma: 1) *an intentionally obscure statement (as a riddle or complex meta-phor) that depends for full comprehension on the alertness and ingenuity of the hearer or reader.*

2) *an inexplicable circumstance, event or occurrence.*

3) *often: one that exhibits an incomprehensible mixture of opposed qualities.*

CHAPTER SIXTEEN

THOUGHTS ON THE NATURAL HISTORY OF EDEN

I n the late 1960s, while a professor at Kansas Wesleyan, I would drive around Saline County looking at rural property for a small homestead, hoping to round out what I considered the perfect life of teaching at a small liberal arts college, coaching track, and raising the three children my wife and I had planned and brought into the world.

On several occasions I found myself parked a few miles south of town by an old iron bridge over the Smoky Hill River. On the east side was a high bank that overlooked the river and a beautiful floodplain opposite. The strip along the high bank most attracted me. It was a strip that had once been broken and farmed, but had since returned to native grassland. Two ravines, one major and one minor, cut their way toward the river, and from them spread trees such as burr oak, green ash, black walnut, hackberry, and box elder. There were the usual accompanying poison ivy and grapevines. Gray dogwood and sumac spilled into the prairie.

It was an idyllic spot. My favorite place was a high point where I could look down on a ripple created by an outcropping of Wellington shale. To sit there was an exquisite experience. I am tempted to say that I meditated on the wonders of nature, but I doubt that. I don't know what I thought. I do know that I was always alone. I never felt like having anyone with me. I was not interested in hunting the land's pheasant, quail, cottontails, or squirrels. Nor had I any desire to fish the stream. The place was Eden.

I learned that an older childless couple, Bessie and Loyd Wauhob, owned this little strip. They lived across the river and across the road. I went to see them and expressed my interest in purchasing a small piece nearest the road, maybe three acres. Bessie's dad had told her never to sell any of her farmland, and she had stuck to that. While both Bessie and Loyd agreed with my assessment that the land was so erodible it had to be abandoned as crop ground, they remained reluctant to sell. They were not the kind of people one should push. But I returned to visit with them a few times and offered as much as $1,000 an acre for three acres. They finally agreed to sell, but protested that $1,000 an acre was too much and that I need pay only $500. A deal was struck.

Of course, I wanted to start building a house and raising the children in the country right away. My wife agreed. The permit office asked to see my plans. There was a tablet on the counter on which I drew the entire perimeter freehand and added a couple of doors. I declared, "Here are my plans." They were approved.

For several months we did nothing to the place, for we lacked the money to start. My family and I frequently came out to picnic or walk around The Land. (That is what we called it and how The Land Institute got its name.) We decided to put the house perpendicular to the setting on the winter solstice and more or less parallel to a slope toward the river north of the ravine, from which a poor-quality coal had been mined more than a half century before. I built a construction shack out of damaged dormitory doors. I added a $25 lab bench from Wesleyan's old science building. Water came from a well my geologist friend Nick Fent drilled across the road. Power and water lines went to the shack in the same trench.

The International Harvester dealer rented me an industrial tractor with a backhoe and a front-end loader at $5 per tractor-hour. I went to work digging the basement, which was to be a walkout affair, filled at the back but mostly open on three sides. After more or less mastering the backhoe, I dug the ditch for the lateral field and the hole for the septic tank. I dug the footing for the house—two to three feet wide and about that deep. Miscellaneous pieces of iron, including old bicycles and tricycles, reinforced the concrete walls poured mostly from self-built forms, intended to be ten inches thick. (I measured one recently, and it is closer to eleven inches.) The unconventional construction included local trees for beams and inside panels.

For rest, I often ambled over to the river, to the same spot where I had stood before construction began, and looked down and out and around. Nothing below had changed. The Wellington shale still generated the ripple. The rustic iron bridge, which should never have been a part of Eden but somehow was, still spanned the river. The fields across the water were the same. To my left remained the large woody ravine. But Eden was gone. I tried to bring it back by opening and shutting my eyes, to image what it was, but it never returned or even came close. Apparently, the very exercise of what makes us human, in that place, drove what some would call "the spirit" away from me. A philosopher might call it a phenomenological experience. I thought of the biblical meaning of the angel with the flaming sword who denies access to Eden.

Arthur Zajonc's book *Catching the Light* helped me understand that the very design of the experiment in trying to determine whether light is a wave or particle determines whether one will perceive a wave or a particle. The design and the perception are a product of our cognition. There is a drawing, an optical illusion, that sometimes appears as a beautiful young woman, and at other times as an old woman. You can't see both at once. In the case of wave versus particle, or beautiful maiden or old woman, it goes back and forth. For me, and my place of Eden, a cognitive switch had been thrown that has never been thrown back.

One interpretation of Genesis may be that our fallen condition comes from insisting that we participate in the Creation. Because I participated in the Creation as a *technological creature*, I had destroyed something whole, which is to say, holy. My family and I, like all others, wanted a home. Who can argue against that motivation? Had I met the shelter need in a minimal sort of way, would Eden have remained? We'll never know. I do know that my perceived need at the time was probably not a real need. We could have continued to live in town. My perceived need was determined more by culture and desire to live in the country than by necessity.

Bessie and Loyd continued to sell us adjacent land as we could afford it until we eventually owned twenty-eight acres. Back from the river now grow various trees bearing organic fruit—beautiful cherries and pears, and also wormy apples. The deer and wild turkey have increased since the early days. The habitat is still safe for them as well as bobcats, quail, pheasants, and nonverbal serpents, not at all tempting us, and less onerous and certainly less toxic than the poison ivy along the big ravine. Nearly four decades have passed since I first experienced Eden there and lost it. Now under the river bluff trees less than a hundred feet away are a sandbox, a playhouse, and a merry-go-round for grandchildren. A sweat-of-the-brow flower and vegetable garden grows over the lateral field, and the shade and pleasing forms of native and exotic trees planted as saplings early on—all watered from the well Nick Fent drilled—bring delight to this participant in the Creation who sometimes thinks the loss of Eden was a bargain.

CODA

We shall not cease from exploration
And the end of all our exploring
Will be to arrive where we started
And know the place for the first time
                    —from "Little Gidding" by T.S. Eliot

The idea of nature as measure is old. As Wendell Berry has helped us understand, it has been carried as a succession in the common culture, but in the formal culture only as a series. For example, from a survey of the writings of British poets up to and including Alexander Pope, several looked to nature as having practical value. After Pope, the thought went underground, and when it resurfaced in the formal culture it was among scientists, beginning with Liberty Hyde Bailey and followed by Sir Albert Howard, J. Russell Smith, and a few other leaders, including Hugh H. Bennett, who in 1935 became the founding chief of what eventually became the Soil Conservation Service. Here this notion, carried as a succession, maybe for ten or twelve millennia, in the common culture had every chance to start the succession in the formal culture. The SCS had ample funding, and Bennett relentlessly emphasized the need to learn from nature's arrangements and apply that knowledge to agriculture from top to bottom. But the SCS failed to stop soil erosion beyond replacement levels for one reason: our grain-producing fields, which occupy 70 percent of agricultural lands, lacked perennial roots. Stuck with annuals, usually grown in monocultures, farmers and SCS workers, especially, have faced the challenge of Sisyphus. A primary purpose of this book has been to argue for the necessity and now the possibility of making the perennial grains available for agronomists and ecologists to bring the processes of the wild to the farm. The idea of nature as measure can then persist as a succession in the formal culture for the first time in agricultural history.

With the explosion of the human population and wild land brought into production, added to the ancient need to solve the problem of agriculture is the need to save what is left of the wild biodiversity throughout the ecosphere. Growing numbers of people beyond ecologists more than "like" biodiversity. They increasingly count wild nature as something close to *sacred*. Saving the wild at the ecosystem or even the species level is perceived now as necessary for stability of the ecosphere and welfare of all within it, including us humans. Agriculture, on the other hand, may be considered necessary, but not "sacred." In fact, many who survey the consequences of agriculture, which can include mining fossil water for irrigation, soil salting, and desertification, as well as soil erosion and loss of biodiversity, regard agriculture as *profane*. We have a dualism: conservation at the cost of production and production at the cost of conservation. This logic has led to advocating higher production on the "profane" acres with more chemical fertilizer, more pesticides, and more irrigation water poured on a landscape that, as one ecologist put it to me, "is already screwed up."

There is more to a future agriculture than merely adopting the perennial habit, though that will yield the largest benefit right off. A shift in our thinking to the ecological end of the spectrum will be necessary. This won't be easy, primarily because the ecosystem as something to study and reflect upon has low standing among biologists and funding agencies. The early architects of modern scientific thought emphasized breaking a problem down into its parts, and in the words of Richard Levins and Richard Lewontin, placed priority on part over whole. As scientists became increasingly reductive, there arose systems theory, something of an improvement, but still another form of reductionism. The ecosystem was still regarded as a container. Emergent properties were largely ignored. In our time, especially since James Watson and Francis Crick's discovery of the structure of DNA in 1953, emphasis on the molecular end of the spectrum has accelerated. That pursuit shouldn't stop. Discoveries made in molecular biology and related side branches are sure to be useful to this new agriculture— but in the service of the ecosystem concept! With the perennial grain crops coming online, agricultural researchers will need ecologists and evolutionary biologists to help apply the great insights gained of how nature's perennial-dominated ecosystems work.

When we set out to look at the structure of the world and rank the levels in a structural hierarchy—think now of Chinese boxes or Russian dolls—we see that right below the ecosphere are ecosystems, the level we must manage for food production. By starting with the ecosphere we gain a sense of respect for matters beyond and larger than people, in five important ways outlined by Stan Rowe: (1) It holds precedence in *time*, since it was here long before we were. (2) It holds precedence in *inclusiveness*. We are embedded within it. (3) It holds precedence in *complexity* of organization, since it is far more complex than we are. (4) It is more *creative*. It beats us in evolutionary creativity and stands as the source of biotic activity. (5) Finally, it has precedence in *diversity*. As he said, nothing is more important than to comprehend the overarching supraorganismic reality, the ecosphere—not biosphere with its bio bias. Recognizing this makes the task of ecology central rather than peripheral. The next level down brings us to the genius of each place, the ecosphere's ecosystems. For agriculture this needs to be the focus of prime importance from here on.

The virtues of local food, organic food, and crop diversity movements, with emphasis on *good* food, have been articulated and acted upon well. The people in these movements are my friends. I am a proud part of that membership. But I am reminded of a 1972 Barry Commoner article in *Harper's*, "Motherhood in Stockholm." There is a section in that article, "Painting the Little Bird." The *New Yorker* cartoon that it is based on shows a spinster lady copying in an art museum from a large painting of massive destruction except for up in a corner, where there is blue sky and a little bird. The lady filled her own canvas by painting just that section of the little bird in the blue sky. Likewise, we mostly ignore the soil lost and the water contaminated because of 70 percent of our farm acreage being devoted to annual grains. On top of this are annual farming's costly inputs from nonrenewable energy and the attendant social consequences.

Two final considerations need mention: the population-consumption problem and the economic system. First, there are too many people, and there are too many things that we want. Herman Daly put the population-consumption problem under one umbrella—population. There is the human population, but there is also the population of deep freezers, the population of houses, the population of pop-up toasters, the population of automobiles. All members of these populations occupy space, and all are dissipative structures, which is to say, all are subject to the second law of thermodynamics, the entropy law. We need to reduce the numbers of many of these populations, including humans. That's number one.

Secondly, we now have to require ourselves to do something the bulk of humanity has not done. All of us creatures are carbon-based and need energy-rich carbon to live, to fuel our bodies, and meet our clothing and shelter needs. Where need meets desire is the problem. In aggregate we humans have shown no more restraint in controlling our numbers and appetites than bacteria on a petri dish of sugar or fruit flies on bananas in a flask or a deer population without predators. We have to lower our population.

Along the way from crude forms of exchange to sophisticated markets, we humans, good at creating abstractions, invented a sophisticated economic system more or less ignorant that growth was the problem. It wasn't planned in the usual sense of that word, but rather self-organized. John Gowdy and Carl McDaniel have stated the problem well:

> The self-organizing principles of markets that have emerged in human cultures over the past 10,000 years are inherently in conflict with the self-organizing principles of ecosystems that have evolved over the past three and a half billion years. The rules governing the dynamics of ecosystems, within which all human activity takes place, are ultimately a function of biological laws, not a function of human-created economic systems. The conflict between these systems is illustrated by the fact that economic . . . environmental indicators have exhibited negative trends.

So, there is more on the line than an abundant and ensured food supply. We have to think about the end of economic and population growth as we know it. We need a nonmaterial growth model, one that is resilient and socially just. So where do we look? Maybe in a seemingly unlikely place—where the split between humans and nature first happened, with agriculture, to The Fall. How might this happen? As the perennial grains come online as a compelling alternative to their annual relatives, with the ecosystem concept firmly in mind, we will seriously begin to assemble them in mixtures, maybe two species at first. As this process is under way, at some point we will observe a quantum leap as the processes of the wild return to the landscape, this time on our farms. Countless efficiencies inherent in the natural integrities common to wild ecosystems, but long absent on these landscapes, will emerge. Is it too outlandish to wonder if, as a new agricultural economy becomes more like nature's economy, our minds will expand? Perhaps all of this may become a new source of metaphors and generally increase our imagination. We will be featuring nature's wisdom over human cleverness—the Tree of Life over the Tree of Knowledge—and have a chance to make T. S. Eliot's "Little Gidding" prophetic.

We must consider more than assuring abundant food. We have the means, if not yet the will, to deal with the population problem. But what are we to do about our species' version of petri dish economics—growth for the sake of growth? We must look to nature's ancient but real economies, which recycle material and run on contemporary sunlight.

We've made gains during these 10,000 years of deficit spending of ecological capital, gains we don't want to lose. We have insight into our stellar origins, reasonable explanations of how a creaturely life could have arisen here, how, as a biologist and Nobel laureate, the late George Wald put it, the "ancient seas set the pattern of ions in our blood, and how the ancient atmospheres molded our metabolism." We products of the simian line know about the speed of light and equations of Newton and Einstein. We know about plate tectonics and continental drift and more, much more, about how the ecosphere works. Literature, art and architecture, history, music, and philosophy have flourished. But these gains after millennia of gathering and hunting were made possible because of our relentless drawing down of pools of energy-rich carbon. Soil erosion and salting, destruction of countless fellow species, are in the wake.

Increasingly, as T. S. Eliot said, we are knowing this place for the first time. We stand at a pivot point. It is redemption time, and we have to redeem ourselves as we redeem agriculture. By starting where our split with nature began, we can build an agriculture more like the ecosystems that shaped us, thereby preserving ecological capital, the stuff of which we are made, and guaranteeing ourselves food for the journey ahead.

ACKNOWLEDGMENTS

My brother, Elmer Jackson (b. 1919), in one of our spirited discussions a few years ago, blurted out in a tone of frustration, "You're always quoting someone else. Don't you have a mind of your own?" I have pondered the insight behind that outburst and conclude that my answer is "No." I don't have a mind of my own, as the reader of this volume will see. My debt to others is total. A major challenge, as I typed this manuscript, was to avoid plagiarizing the many well-said thoughts, ideas, and insights that have come my way through conversations and in writing. Along the way I began to write down names of people to whom I owed a debt. As the list grew longer, I grew more worried that I had left someone out, not just some *one*, but lots of someones. Even though this is not a scholarly work, but amounts now to an essay since I cut it in half, the need to acknowledge those to whom I owe a debt did not decline proportionally.

The necessary scholarly treatments have already been done by such luminaries as Hans Jenny (*The Soil Resource*, 1980), Daniel J. Hillel (*Out of the Earth: Civilization and the Life of the Soil*, 1991), David R. Montgomery (*Dirt: The Erosion of Civilizations*, 2007), John Reganold and many others. All have deepened my understanding and appreciation of soil under threat beyond my poor powers to add one iota.

Speed-dial friends and mentors include Bill Vitek, former Land Institute board chairs Don Worster and Conn Nugent, and current chair Angus Wright. Other compass correctors include Herman Daly, Paul Heltne, David Orr, Ted Lefroy, Fred Kirschenmann, and David Ehrenfeld. Then there are those who have devoted decades of their productive lives to the necessary massive salvage operation: Agrarians Maury Telleen, Gene Logsdon, and David Kline. Doug and Kris Tompkins are protectors of wild ecosystems in Chile and Argentina. Ken Whealy, co-founder of Seed Savers Exchange, is an authority on the complications as well as the possibilities for saving heirloom varieties.

I must also acknowledge the intellectual contributors of the Mt. Sky gathering every two years in Montana, and its host, Charlie Sing.

Aside from those already mentioned, I must thank Peggy Waggoner, who did early germplasm development on intermediate wheatgrass. Those at work on plant breeding here at The Land Institute—Stan Cox, Lee DeHaan, David Van Tassel, and Cindy Cox—and soil ecologist Jerry Glover are the backbone of our research agenda. Their words are throughout this book, especially in chapters 10 and 11. A few of the contributions of the late Marty Bender, my longest standing colleague here at The Land, are spelled out in the text. He died too young, and we miss him.

The heavy lifting quickly and professionally accomplished by Darlene Wolf during the preparation of this manuscript while handling her countless other duties as my special assistant is appreciated more than I can express.

Trish Hoard, Deborah Rich, Scott Bontz, Carrie Carpenter, and Duane Schrag provided invaluable editorial assistance, as did Laura Krinock and Jean Kozubowski, early organizers of my chaos, Freddie Smith more recently, and Elizabeth Mathews, Laura Mazer, and Jack Shoemaker of Counterpoint. And thanks to photographers Jim Richardson and Steve Renich for their photographs.

Two good friends lost within a year and a half left me with a sadness measured by their importance to my thinking. Strachan Donnelley, Land Institute board member, philosopher, and founder of the Center for Humans and Nature, is one. Charles Washburn, metallurgist by training, was a rigorous numbersmith and an articulate critic, and over the last thirty-eight years was always available when I needed clarity. They are in this book.

The long and meaty conversations I have with my ecologist daughter Laura Jackson and her husband Kamyar Enshayan are at once delightful and informative. The volume edited by her mother, Dana Jackson, and her—*The Farm as Natural Habitat: Reconnecting Food Systems with Ecosystems*—is useful for informing current farmers of the possibilities when we look to nature.

Over the past fifteen years I have had almost daily conversations with my good friend Ken Warren, geologist, paleobotanist, and managing director here at The Land Institute. It is hard to leave his presence without having learned something. Sometimes it is essential to the daily and future operation of the place, but just as often an interesting account of a substansive nature. Shallow and conventional he is not, as he manages the flame of this place to ensure good work.

Finally, my indefatigable wife, Joan, is always a willing and helpful critic. She is more than director of institutional advancement here at The Land Institute. She is governor of my impulsive behavior. Her patient and cheerful demeanor beyond my understanding surely must be due to her strong sense of the aesthetic that daily blooms. Dedicating this volume to her is a very small token of my gratitude.

REFERENCES
AND NOTES

.

Preface

ix   The original title of Alexander Pope's poem was *An Epistle to the Right Honourable Richard Earl of Burlington, Occasioned by his Publishing Palladio's Designs in the Bathes, Arches, and Theatre's of Ancient Rome.* This is *Epistle IV*, the last one. The main idea of the poem is to follow nature.

PART I: SOME HISTORY AND ASSUMPTIONS
Chapter One—Introduction

3   Nicole Krauss, *The History of Love* (W.W. Norton & Company, 2005).

4   The U.N. Environment Program's first *Global Environment Out-
        look Year Book* was released in 2003. The program executive direc-
        tor, Klaus Toepfer, noted that the dead zone problem was likely to
        rapidly escalate. He stated that there are 146 dead zones, most of
        which are in Europe and the east coast of the U.S. The most infa-
        mous is at the end of the Mississippi River, due to fertilizer from
        the farm fields of the Midwest.

7   Joel Cohen, *How Many People Can the Earth Support?* (W.W.
        Norton & Company, 1996).

8   National Academy of Sciences, July 2002.

8   David Pimentel, et al., "Environmental and Economic Costs of
        Soil Erosion and Conservation Benefits," *Science* 267, no. 5201
        (1995): 1117–1123.

9   Camille Parmesan, "Ecological and Evolutionary Responses to
        Recent Climate Change," *Annual Review of Ecology, Evolution,
        and Systematics* 37 (2006): 637–639.

10   Kathleen Raine, *Defining the Times: Essays on Auden and Eliot*
        (London: Enitharmon Press, 2002).

15   Sir Albert Howard, *An Agricultural Testament* (Rodale Press,
        1972).

16   Rene Descartes, *Meditations on First Philosophy* (CreateSpace,
        2009).

18   Fred Hoyle, *Of Men and Galaxies* (Prometheus Books, 2005).

Chapter Two—One Man's Education

29   Aldo Leopold, *A Sand County Almanac: With other essays on con-
        servation from Round River* (Oxford University Press, 1966).

31   *Figure* 4. Courtesy of Aldo Leopold Legacy Center, Baraboo, WI.

34   The second edition of *New Roots for Agriculture* was published by the University of Nebraska Press in 1985.

37   Parts of this piece on Wendell Berry first appeared in *Wendell Berry: Life and Work (Culture of the Land)*, edited by Jason Peters (University of Kentucky Press, 2007).

40   "Nature as the Measure for a Sustainable Agriculture" is a chapter in *Ecology, Economics, and Ethics: The Broken Circle*, edited by F. Herbert Borman and Stephen R. Kellert (Yale University Press, 1961).

42   Liberty Hyde Bailey, *The Outlook to Nature* (Forgotten Books, 2010). Also see *The Holy Earth* (University of Michigan Library, 2009).

42   Albert Howard, *An Agricultural Testament* (Rodale Press, 1972).

45–46   Much, if not most, of this history about Arnold Schultz is drawn from his writings under the title "What Is Ecosystemology?" and has never been published. It can be found online at http://nature.berkeley.edu/classes/ecosystemology.

47   E. P. Odum, *Fundamentals of Ecology*, 3rd ed. (WB Saunders Co., 1971).

49   I recommend especially Stan Rowe's two books, *Home Place: Essays on Ecology* (1990) and *Earth Alive: Essays on Ecology* (2006), both by NeWest Press, the second published after his death.

52   James Lovelock, *Gaia: A New Look at Life on Earth*, 3rd ed. (Oxford University Press, 2000).

Chapter Three—Earth Is Alive

64   See page 145 of *Earth Alive* in first paragraph entitled "Considering Life" in context for quote at head of this chapter.

64   Gregory Bateson, *Steps to an Ecology of Mind* (Chandler Publishing Company, 1972). It was based on the 19th Annual Korzybski Memorial Lecture delivered January 9, 1970.

69   William C. Noll, "Environment and physiological activities of winter wheat and prairie during a year of extreme drought" (January 1, 1936). *ETD collection for University of Nebraska–Lincoln*. Paper AAIDP13879. http://digitalcommons.unl.edu/dissertations/AAIDP13879.

Chapter Four—The 3.45-Billion-Year-Old Imperative and the Five Pools

79   Vaclav Smil, *Nature* (July 29, 1999): 415.

81   Sources for these figures derived from United States Census Bureau, which gives daily estimates of both world and U.S. population. Oil curve derived from various depletion analysts. A good beginning point is with M. King Hubbert, who, as early as 1956, created models. Given the power of exponential growth, doubling and tripling, the estimates of unknown resources change the year of peak oil only slightly.

82   Global Perspective Studies Unit, Food and Agriculture Organization of the United Nations, *World Agriculture: Towards 2030/2050, Interim Report*, FAO, Rome, 2006.

82   Energy Information Administration, "International Energy Outlook 2007," U.S. Department of Energy Report No. DOE/EIA-0484, May 2007.

84   David Tilman, et al., "Agricultural sustainability and intensive production practices," *Nature* 418 (August 8, 2002).

Chapter Five—The Rise of Technological Fundamentalism

85–86   Robert W. Fuller, R. B. Brownlee, and L. Baker, *Elements of Physics* (Allyn and Bacon, 1947). Opening paragraph.

88   William Stanley Jevons, *The Coal Question* (General Books LLC, 2009).

88   *The Oxford Book of Aphorisms*, Oxford University Press (1983), attributes "Invention is the mother of necessity" to Thorstein Veblen.

Given the title of this chapter and even the last, it is worth quoting a larger context of Wallace Stegner. In his 1953 book, *Beyond the Hundredth Meridian*, he describes the breakdown of American Indian culture as follows:

"However sympathetically or even sentimentally a white American viewed the Indian, the industrial culture was certain to eat away at the tribal cultures like lye. One's attitude might vary, but the fact went on regardless. What destroyed the Indian was not primarily political greed, land hunger, or military power, not the white man's germs or the white man's rum. What destroyed him was the manufactured products of a culture, iron and steel, guns, needles, woolen cloth, *things that once possessed could not be done without.* [Italics added.]

"It was not the continuity of the Indian race that failed, what failed was the continuity of the diverse tribal cultures. These exist now only in scattered, degenerated reservation fragments among such notably resistant peoples as the Pueblo and Navajo of the final persistent Indian Country. And here what has protected them is aridity, the difficulties in the way of dense white settlement, the accident of relative isolation, as much as the stability of their own institutions. Even here a Hopi dancer with tortoise shells on his calves and turquoise on his neck and wrists and a kirtle of fine traditional weave around his loins may wear down his back as an amulet a nickel-plated Ingersoll watch, or a Purple Heart medal won in a white man's war. Even here, in Monument Valley where not one Navajo in ten speaks any English, squaws may herd their sheep through the shadscale and rabbitbrush in brown and white saddle shoes and Hollywood sunglasses, or gather under a juniper for gossip and bubblegum. The lye still corrodes even the resistant cultures."

92   Donella Meadows, "Leverage Points: Places to Intervene in a System," *Whole Earth*, Winter 1997.

93   Jay W. Forrester, "Counterintuitive Behavior of Social Systems." This paper, copyrighted 1971, was based on testimony for the Subcommittee on Urban Growth of the Committee on Banking and Currency, U.S. House of Representatives, on October 7, 1970. It is mentioned here partly because of how early in the so-called environmental movement the problem of growth was mentioned to the U.S. Congress.

93   Morris Berman, *Dark Ages America: The Final Phase of Empire* (W.W. Norton, 2006).

PART II: LOSSES
Chapter Six—The Most Serious Loss of All

100   Carl O. Sauer, "Agency of Man on Earth." Excerpt from *Man's*

Role in Changing the Face of the Earth (University of Chicago Press, 1956).

101   W. C. Lowdermilk's "Conquest of Land through 7,000 Years" is a report of a travel/study effort in 1938 and 1939. It is available as a reprint from the Natural Resources Conservation Service of the U.S. Department of Agriculture without charge.

101   *The Epic of Gilgamesh* can be found in several places, including *World Mythology: An Anthology of the Great Myths and Epics* (Lincolnwood, IL: Donna Rosenberg National Textbook Company).

105   C. N. Runnels, "A Greek Countryside: The Southern Argolid from Prehistory to the Present Day," *Scientific American*, March 1995.

107   Daniel J. Hillel, *Out of the Earth: Civilization and the Life of the Soil* (University of California Press, 1992).

107   David R. Montgomery, *Dirt: The Erosion of Civilizations* (University of California Press, 2007).

107   Z. Naveh and P. Kutiel, *Changes in the Mediterranean Vegetation of Israel*.

107   F. Taylor, "The destruction of the soil in Palestine," *Bulletin of the Soil Conservation Board of Palestine*, 1946.

107   C. Vita-Finci, *The Mediterranean Valleys: Geological Changes in Historical Times* (Cambridge University Press, 1969).

107   R. Lal, "Soil Erosion and Land Degradation: The Global Risks," R. Lal and B. A. Stewart (eds.), *Advances in Soil Science 11: Soil Degradation* (New York: Springer-Verlag, 1990).

107   A. Fitzgerald, "Mining Iowa's Gold," *Des Moines Register*, September 6, 1998.

107   Lester R. Brown, "The global loss of topsoil," *Journal of Soil and Water Conservation* 39 (1984): 162–165.

107   Currently more than two billion hectares of soil have been degraded worldwide by human activities. See especially S. Wood, K. Sebastian, and S. J. Scherr, *Pilot analysis of global ecosystems: Agroecosystems* (Washington DC: IFPRI and World Resources Institute, 2000). Also see UNEP, *GEO-3: Past, Present and Future Perspectives* (London: Earthscan, 2002).

107   The percentage of degraded land, based on worldwide estimates, ranges from 15 to 38 percent of cropland degraded by agricultural activities over the last fifty years. See S. J. Scherr and D. Yadav, *Land degradation in the developing world: implications for food, agriculture, and the environment to 2020* (Washington DC: IFPRI, 1996).

107   Dr. William F. Ruddiman, author of a marvelous book, *Plows, Plagues & Petroleum* (Princeton University Press, 2005), has a chapter entitled "Early Agriculture and Civilization" that I wish I could lift for this volume. He describes the spread of both crops and livestock, their current geographic distribution, as well as the spread of technology.

109   There were 1.6 million applications by 1934 and some 270 million acres passed into private hands. See the National Archives and Records Administration (NARA). Archives.gov/education/lesson/homestead-act.

110   Wendell Berry, *The Unsettling of America* (Sierra Club Books, 1977).

117   Bud Moore, *The Lochsa Story: Land Ethics in the Bitterroot Mountains* (Mountain Press Publishing Company, 1996).

119   Wellington Brink, *Big Hugh: The Father of Soil Conservation*

(McMillan, 1951). Also in 1951 is a piece by W. Brink in *The Land* X, no. 3.

122   Russell Lord, "The Whole Approach," *The Land and Land News* IX (Winter 1953): 4. This issue was a special review and preview for 1941–1953.

127   Summary of Appraisal, Parts I and II and Program Report, Soil and Water Resources Conservation Act (RCA). Review Draft, 1980. USDA. Also Review Draft Part I.

129   Jerry T. Harlow, *History of Soil Conservation Service National Resource Inventories*, USDA NRCS paper (1994).

129   Report from the Economic Research Service, "The Conservation Reserve Program: Economic Implications for Rural America," September 2004.

Chapter Seven—Chemicals on the Landscape

132   Terry Collins, "Toward Sustainable Chemistry," *Science* 291, no. 5501 (2001): 48–49.

132   D. Pimentel (ed.), *Handbook of Pest Management in Agriculture* I–III, 2nd ed. (Boca Raton, FL: CRC Press, 1991).

133   David Tilman, et al., "Agricultural sustainability and intensive production practices," *Nature* 418 (August 8, 2002).

133   E. M. Bell, et al., "A case control study of pesticides and fetal death due to congenital abnormalities," *Epidemiology* 12 (2001): 148–156.

133   L. Hardel and M. Eriksson, "A case-control study of non-Hodgkin lymphoma and exposure to pesticides," *Cancer* 85 (1999): 1353–1360.

133   C. Lu, et al., "Pesticide exposure of children in an Agricultural

Community: Evidence of household proximity to farmland and take-home exposure pathways," *Environmental Research* 84 (2000): 290–302.

134   A. Blair, et al., "Clues to cancer etiology from studies of farmers," *Scandinavian Journal of Work, Environment & Health* 18 (1992): 209–302.

134   A. Blair and S. H. Zahm, "Cancer among farmers," *Occupational Medicine* 6, no. 3 (1991): 335.

134   S. H. Zahm, et al., *Cancer Research* (Suppl.) 52 (1992): 5485s–5488s.

134   R. Repetto and S. Baliga, *Pesticides and the Immune System: The Public Health Risks* (Washington DC: World Resource Institute, 1996).

Chapter Nine—Global Warming:
How a Little Change Can Mean a Lot

140   Richard Levins. Quotes here are from an unpublished manuscript; see Richard Lewontin and Richard Levins "Schmalhausen's Law," *Capitalism Nature Socialism* 11, no. 4 (2000): 103–108; also see Richard Levins, et al., "Variability and Vulnerability at the Ecological Level: Implications for Understanding the Social Determinants of Health," *American Journal of Public Health* 92, no. 11 (November 2002); also see "Musings on human ecology, public health, philosophy of science, socialism, capitalism, Cuba, ants, current events . . ." http://richardlevins.com/2009/10/28/schmal-hausens-law.html.

PART III: REVERSING THE DAMAGE
Chapter Ten—Consulting the Genius

147 William L. Thomas Jr. (ed.), *Man's Role in Changing the Face of the Earth* (University of Chicago Press, 1956).

147 Chuck Washburn, personal communication.

148 John J. Ewel, M. Mazzarino, and C. Berish, "Tropical soil fertility changes under monocultures and successional communities of different structure," *Ecological Applications* 1, no. 3 (1991): 289–302.

152 Christopher B. Field, "Sharing the Garden," *Science* 294 (December 21, 2001).

160 John L. Harper, *Population Biology of Plants* (Academic Press, 1981): 656.

160–161 L. T. Evans, *Crop Physiology: Some Case Histories* (Cambridge University Press, 1978): 334, 331.

171 William T. Coyle, "Next-Generation Biofuels: Near-Term Challenges and Implications for Agriculture," *USDA: A Report from the Economic Research Service*, May 2010.

There have been countless arguments about a massive biofuels program. Rather than argue about energy balances for making ethanol out of plants and how much water that will take, it is far more important to look at our current demand. For the United States, my late friend Charles Washburn began calculations by assuming that we use around 20 million barrels of oil per day, or around 7.3 billion barrels per year. During that year the average American eats the energy equivalent of about 26 gallons of gasoline. Let's say we stop eating for a year and turn all of that food into ethanol. Even though the energy value of ethanol per

gallon is less than gas or diesel, let's assume a little over 2 gallons a month per person. That's for everything that now uses
oil: buses, trains, airplanes, letter carriers, and so on. In addition
to not eating for a year, let's harvest from our fields all of the
corn stalks, all the wheat straw, and all the misnamed "waste."
That should give us another 2 gallons a month. Where else can
we look? Let's stop all food exports to foreign markets. This will
give us another 1.5 to 2 more gallons per month, for about 6
total. A family of four has 24 gallons per month.

Charles Washburn calculated that a combination of our planet's rice, wheat, corn, and soybean crops turned into ethanol
and biodiesel would meet only 14 percent of the U.S. demand
for vehicle fuel. This biomass includes energy-rich carbon that
microbes and invertebrates need for food. At the soil community
level, it makes possible those mineralization processes necessary
to make elements available for future growth. We'd better hope
there is enough energy-rich carbon below the surface, exuding
from roots or dead organisms, to feed those critters. By pulling off "waste" beyond nutrients removed from food, we have
accentuated the soil mining process, and so we'll have to add
nutrients, and soil erosion is sure to accelerate.

172   James N. Galloway and Ellis B. Cowling, "Reactive Nitrogen and
        the World: 200 Years of Change," *Ambio: A Journal of the Human
        Environment* 31, no. 2 (March 2002).

173   David Tilman, et al., "Carbon-Negative Biofuels from Low-Input
        High-Diversity Grassland Biomass," *Science* 314 (2006): 1598.
        The following papers authored by Land Institute scientists rep-
        resent further background and recent developments:

T. S. Cox, D. L. Van Tassel, C. M. Cox, L. R. DeHaan, 2010. "Progress in breeding perennial grains." *Crop and Pasture Science.* 61, 1–9.

L. R. DeHaan, D. L. Van Tassel, T. S. Cox, 2005. "Perennial grain crops: A synthesis of ecology and plant breeding." *Renewable Agriculture and Food Systems.* 20(1); 5–24.

Jerry D. Glover, John P. Reganold. "Perennial Grains: Food Security for the Future." *Issues in Science and Technology.* Winter 2010.

Jerry D. Glover, et al., 2010 "Increased Food and Ecosystem Security via Perennial Grains." *Science.* 328: 1638–1639.

D. L. Van Tassel, L. R. DeHaan, T. S. Cox, "Missing domesticated plant forms: can artificial selection fill the gap?" *Evolutionary Applications* (vol. 3: 434–452).

Chapter Twelve—An Appeal to the Russians

192   Jacob N. Barney and Thomas H. Whitlow, "A unifying framework for biological invasions: the state factor model," *Biological Invasions* 10 (2008): 259–272.

193   Gary Paul Nabhan, *Where Our Food Comes From: Retracing Nikolai Vavilov's Quest to End Famine* (A Shearwater Book published by Island Press, 2009).

195   N. V. Tsitsin, "Remote hybridization as a method of creating new species and varieties of plants," *Euphytica* 14 (1965): 326–330.

197   N. I. Vavilov, "The Law of Homologous Series in Variation," *Journal of Genetics* 12, no. 1 (April 1922).

Chapter Thirteen—Were Ants the First Agriculturists?

201   H. E. Jacobs, *Six Thousand Years of Bread: Its Holy and Unholy History* (Doubleday, Doran and Co., 1944).

202–203   Darwin reading from above reference.

204   Robert Orr Whyte, "The Botanical Neolithic Revolution," *Human Ecology* 5, no. 3 (September 1977): 209–222.

PART IV: ANALYZING THE RESISTANCE
Chapter Fourteen—Analyzing the Resistance

209   Ernst Mayr, *The Growth of Biological Thought: Diversity, Evolution, and Inheritance* (Cambridge, MA: The Belknap Press of Harvard University Press, 1982).

210   Shankar Vedantam, "Persistence of Myths Could Alter Public Policy Approach," *Washington Post*, September 4, 2007. This article cites the research in the July 18, 2006, Pew Global Attitudes Project.

212   Morris Berman, *Dark Ages America: The Final Phase of Empire* (New York: W.W. Norton, 2006).

212   James Surowiecki, "Fuel for Thought," *The New Yorker*, July 23, 2007.

213   Dan Luten, "Empty Land, Full Land, Poor Folk, Rich Folk," *Yearbook of the Association of Pacific Coast Geographers* 31 (1969).

213   Wendell Berry, *The Unsettling of America: Culture and Agriculture* (San Francisco: Sierra Club, 1977).

PART V: AWAY FROM THE EXTRACTIVE ECONOMY
Chapter Fifteen—Away from the Extractive Economy

218   Elliot West, *The Contested Plains: Indians, Goldseekers, and the Rush to Colorado* (University of Kansas Press, 1998).

232   Patrick Kavanagh, "The Parish and the Universe" 1 (orig. "Mao Tse-tung Unrolls His Mat," in *Kavanagh's Weekly* 7 [May 24, 1952]).

PART VI: AN ENIGMA
Chapter 16—Thoughts on the Natural History of Eden

245   Richard Levins and Richard C. Lewontin, *The Dialectical Biologist* (Cambridge, MA: Harvard University Press, 1985).

245   Stan Rowe, *Home Place: Essays on Ecology* (Edmondton, Alberta: NeWest Press, 1990).

246   Commoner, Barry, "Motherhood in Stockholm," *Harper's Magazine* (June 1972).

247   John M. Gowdy and Carl N. McDaniel, "One world, one experiment: addressing the biodiversity-economics conflict," *Ecological Economics* 15, no. 3 (December 1995): 181–192.

248   T. S. Eliot, *Four Quartets* (San Diego: A Harvest Book—Harcourt, Inc., 1943).

COLOR INSERT

tall wheat? Probably both. Note percent of agriculture landscape devoted to grain. Apple and pear trees limbed up presumably to allow light for growing grain.

Fig. 8　On left in each panel is annual wheat except in summer panel after harvest and before planting. On right in each panel is perennial intermediate wheatgrass.

Fig. 9　Excavation to show differences in root density and depth.

Fig. 10　Field trial of 1,000 intermediate wheatgrass plant genotypes divided into three clonal replicates.

Fig. 11　Intermediate wheatgrass. A typical response to selection of a formerly wild herbaceous species for desirable agronomic traits.

Fig. 12　These regrowing plants are the result of crosses involving two species: annual wheat and perennial intermediate wheatgrass. Domestic wheat will cross with scores of its relatives. Since it is self-fertile (an individual plant will accept its own pollen), anthers must be removed before pollen is shed if egg is to be fertilized by another plant. Sometimes we can overcome a species barrier by what is called embryo rescue. Endosperm is a food source for the embryo. Often when a wide cross is attempted, fertilization may be successful and an embryo formed but the necessary nurturing endosperm will not develop, starving the embryo. Therefore, Land Institute researchers rescue the embryo using tweezers and place it on a nutrient medium in a test tube. If the plant survives, it may be successful in producing endosperm when subsequent backcrosses are made to the parents. In this photograph some are hybrids and others products of hybrid backcrosses to either parent. After harvest, plants that experience regrowth become candidates for further crosses.

Fig. 13   Chromosome "painting" makes it possible to identify which parent each chromosome came from in an earlier cross. Here the red ones are from an original annual wheat, the green from the perennial intermediate wheatgrass. Chromosomes shown here are from a hybrid between the annual and perennial back-crossed to the perennial.

Fig. 14   Tetraploid (four sets of chromosomes) grain sorghum parents (1) were crossed with strains of johnsongrass (2) to produce hybrids (3) that resembled the wild johnsongrass. Hybrids or their perennial offspring were crossed again to grain sorghum, producing diverse backcross populations (4) with "tamer" characteristics. Perennial plants in those populations have been crossed with a more diverse group of grain sorghum strains (5) to produce a second backcross generation from which perennial plants are selected.

Fig. 15   Perennial sorghum breeding nursery.

Fig. 16   Illinois bundleflower, a native nitrogen-fixing legume, for a wild species has high seed yield and good nutritional quality. With aggressive breeding it could be a partial substitute for soybeans.

Fig. 17   Individual bundle of Illinois bundle flower with pods opening. Shatter resistance of seeds is a necessary trait for domestic crops.

Fig. 18   Annual sunflower domestication was accomplished by Native Americans. The wild annual (left) with numerous small heads and seeds was the ancestral type for the form with large head and seeds typical of domestic form today.

Fig. 19   Wild annual sunflower (A) and domesticated form (B) have both been crossed with two wild species: *Helianthus tuberosa* (D) and *H. rigidus* (E). Hybrid gene pool (C) from numerous hybrids between species in turn have been crossed with one another.

Fig. 20   Upland rice in China. Note erosion in background and upland rice plants in foreground.

Fig. 21   Dr. Fengyi Hu showing upland rice plant with perennial rhizomes at the Yunnan Academy of Agricultural Sciences (YAAS).

Fig. 22   Native prairie at The Land Institute (foreground). Research plots (background) of several perennial grain species.

Fig. 23   The native prairie standard or measure with diversity both above and below. Dark soil indicates an abundance of stored carbon, more important in soil than in the gas tank.

Fig. 24   The grasslands evolved to invite fire, and Native Americans used fire for quick greening to lure in the bison. Many ranchers in the Kansas Flint Hills, featuring hundreds of thousands of acres of tallgrass prairie, use fire as a management tool today.

Fig. 25   The bison herd at The Land Institute selectively grazes on the native prairie year-round.

Fig. 26   Cattle, domestic analogs of the bison, on the Kansas Flint Hills are selective grazers and are being considered, along with bison, to be future management tools for perennial grain-producing polycultures.

Fig. 27   The swather and mowing machine are nonselective grazers and may prove to be an effective management tool for perennial grain-producing polycultures.

Fig.28   This team of Belgians is pulling a sickle mower, a nonselective grazing machine using solar-powered muscle.

Fig. 29   Four-horse hitch pulling a fore-cart with a gasoline engine, necessary for a power takeoff for the crimping machine, which squeezes juice out of the hay to speed drying. Note extra energy cost in horsepower and fossil fuel use to speed drying and to maximize harvest.

Fig. 30   One can imagine hay bales becoming a fuel for on-farm mechanical work. If such biofuel can be produced near the farm and restricted to on-farm use, one can imagine such fuel harvest integrated with perennial polyculture grain production. (See pages 171–174.)

Fig. 31   Illinois corn field, June 2008. Note how level the field is.

Fig. 32   Processes such as present in this wild prairie are waiting to be tapped for future use in seed-producing perennial polycultures.